Books by Simon Clarke

Self Help – How to get started.

Two months in America.

How to write and publish your own book for free

Three back cats

Tommy and the never-ending Mountain

This will blow your socks off

This will blow your socks off too

Slick Moaner Adventures

Adventures into Danger #1

Adventures above the Clouds #2

Adventures of this World and Others #3

Contact the author @

SimonAmazingClarke@gmail.com

DISCLAIMER AND TERMS OF USE AGREEMENT

The Author, Editor and Publisher of this book, and the accompanying materials have used their best efforts in preparing this book. The author and publisher make no representation or warranties with respect to the accuracy, applicability, fitness, or completeness of the contents of this book.

Therefore, if you wish to apply ideas contained in this book, you are taking full responsibility for your actions.

The Author, Editor and Publisher disclaim any warranties (express or implied), merchantability, or fitness for any particular purpose. The author and publisher shall in no event be held liable to any party for any direct, indirect, punitive, special, incidental or other consequential damages arising directly or indirectly from any use of this material, which is provided "as is", and without warranties.

This book is © copyrighted by Simon 'Amazing' Clarke. No part of this may be copied, or changed in any format, sold, or used in any way other than what is outlined within this book under any circumstances.

Comments

It is great, as an author, to get comments from people who have read your book. Here is what Kevin M. thought after he had read an early copy.

It is factual and explains the principles well. Almost like the dummies guides, but better as there is some personal and practical feel to it.

David C. was even more inspired.

You don't have to be Einstein to know that Tesla had the right idea 100 years ago (Einstein was a big fan of Nikola Tesla's work.)

The EV revolution has started, but most people haven't noticed.

You don't have to be a spoon whittler to live off the grid.

Wow, now there is some inspiration. I wonder what your thoughts will be once you have read this book.

Index

1	Why renewables	5
2	Electric Cars	9
3	Electric Cars, Part Two	30
4	eBikes, Scooters and Motorbikes	52
5	Houses	57
6	Other Home technologies	65
7	Our Short Term Future	67
8	My Ideal House	75
9	Our Long Term Future	78
10	2019 mid-year update	89
11	Finishing Words	97

Chapter 1, Why Renewables

I want you to imagine this situation. You are on your way home from work. Your fuel gauge is showing less than a quarter of a tank, so you decided to fill up your car. As you get to your local garage, you find that cars are queuing out onto the road. You pass by and continue to your house. You enter your house and realise it is cold. You reach for the first light switch, and nothing happens. You try several others and come to the slow realisation that there must be a power cut of some description. You know you have a couple of torches and, with your gas cooker and gas fire, you can at least cook yourself a meal and be warm in your lounge. You put a pan of water on the stove, turn the knob and, again, nothing happens. There is no hiss of gas. You are totally out of energy. So how did this happen?

We in the western world are so reliant on being provided with services that we have grown to expect them. But there are people who campaign against us having them. In the above scenario, the campaigners have won in the fight against the new nuclear power stations and the tidal energy production. They have stopped the Fracking and stopped us, as a country, importing foreign oil and gas.

So what can we do? Well, currently we have increasing demands as our populations grow year on year. The oil and gas companies help to provide the energy that we demand. I watched a film clip where a large number of activists had got into their canoes to prevent a new oil platform being moved. The problem here is, if there weren't the demand, the energy companies wouldn't need to build these large oil drilling contraptions. I do wonder though, how many of those protesters arrived in the large gas guzzling suvs and 4x4's.

So what can we do? We either need energy, or we freeze and starve.

There is, fortunately, the third option and, due to high levels of technology, everything that we need is here with us today.

The big question is, how much more will it cost for me to live my life without the need to rely on both fossil fuels and without importing energy from other countries?

The real answer is, no more than it currently costs now, and I, and you, might even be better off. I would add to this that with the rate of advancing technologies, you will be able to achieve 100% self-sufficiency and also produce excess energy. This excess can go back into the national grid. This will happen within the next five to ten years.

Look out for a TV program that came out in May 2017 that is presented by Robert Llewellyn. We all know how glitzy and lit up Las Vegas is. I can't imagine what their energy bill is. Actually, I know what it is going to be, zero. How are they achieving this? They are going totally solar and battery power. That is how far technology has moved on in the last few years.

So let's go back to the scenario at the start of this chapter and replay it with the new technology.

I drive home in my car, I notice the massive queues in front of the garage. I've not been to one of those in years since I bought my electric car. I arrive home and plug my car in ready for the following day noticing that the street lights are not working and that most of the houses around me are in darkness with only the odd torchlight and flickering candle showing. I enter my house, and I am met with a lovely warmth. With my cavity wall insulation, low energy loss windows, extra thick roof insulation and well-sealed doors my heating energy usage is very low. I put a pan of water on the stove and turn it on. I notice that the flame is very clear; this indicates that it has a high hydrogen content. I sit down to check my computer, and I see that I have exported today. Not bad, 3kWh of electricity and 0.75 cubic meters of hydrogen. That was after everything that my house had used during the day and after topping up my house battery and hydrogen tank.

We all understand why we need to reduce our reliance on fossil fuels. This book is going to show you how, and with the help of the companies in this book, and of those in your area, and with current government schemes, you can change your life to achieve this. Jack Canfield in his book 'How to get from where you are to where you want to be,' gives us rule number one, 'you are 100% responsible for you'. So if you want to change the world, change your life without compromising your lifestyle today.

Let's start by looking at some new terminology:-

ICE – Internal Combustion Engine, these can be Petrol/Gasoline or Diesel.
EV – Electric Vehicle, usually referring to cars at this time.
HEV and **PHEV**, I will explain later
PV – Photo Voltaic, how light is turned into electricity in Solar Panels.

In this book, we will be covering not only electric and hybrid cars but also solar panels, house batteries and UK grants. We will look at other ways of improving your home's energy efficiency with cavity wall insulation, greater loft insulation. Lower energy bulbs, and how you can get them for free (UK), commercial electric lawn mowers and running your house on Hydrogen. Then we will look forward to seeing what the future holds.

Please remember that these are my comments from the information that I have found. I do not claim to be an expert. Depending on where you live, the laws, technology availability and possible grants etc. will differ. See what is available near you and, if it helps you save money, go for it. As a by-product of us using this new technology, we will all help to do our bit to clean up the environment.

You will learn more as we go on, but I'll tell you one thing, I want an all-electric car, solar panels on my house and a house battery. I want to be able to charge my car up from these batteries and have a sticker on the back of it that says 'This car is powered by sunshine.'

Welcome to your new paradigm.

Chapter 2, Electric Cars

I'll tell you what came into my mind when I heard that they were starting to produce electric cars, a Milk Float. When I was a young lad in the 70's, and 80's there used to be a milkman who made deliveries to our neighbourhood.

The design of the Milk Float was perfect for a door to door type of deliveries. There were four wheels as per a standard car design. There was a seat, steering wheel and pedals, again, all pretty much standard. It had an enclosed cab, but I can't remember there being any doors. Lastly, there was a flatbed on the back with a fibreglass cover over the top to keep the worst of the weather off.

These were electrically powered with a 1960's technology motor and a couple of tons of lead acid batteries. They would require an overnight charge to prepare them for the following day's work. They were a significant advance on what they were designed to replace, namely horse-drawn milk Floats. The electric ones would go further in a day and go faster, never mind the fact that they always did as they were told, didn't require feeding, farriers or vets.

From what I could see they were well designed for the task. The milkmen would pop in and out of them as they made their way through their rounds.

About twenty years ago a guy had an old Range Rover with a worn out engine. From somewhere he bought an old milk float and transferred over the batteries and motor. Range Rovers can carry a lot of weight, but the 1.5 tons of batteries were a struggle to carry. The car was limited to around 20 mph, and it did have a limited range.

With this in mind, I did wonder just how good new electric cars were going to be.

I was certainly impressed when the Nissan Leaf came out around 2013. It was a car that looked like a proper car. It had proper car top speeds. There was no way it was going to be the fastest car on the block, nor the prettiest, but if you wanted a commuter car that would take you from A to B up to 20 miles each way, or 40 miles if you had a charging point at work, then this would work for you. What was more impressive at the time was that you could plug your car into 'public' chargers for FREE!

Although I knew of the Toyota Pries, I didn't know what it was and how it worked. It turns out that it is an HEV.

Now let me explain what some of these new acronyms mean. We all know and understand what I will call 'conventional' cars. This is what we have been driving around for years. They were powered mainly by petrol engines but more recently by diesel ones. The UK and the rest of Europe, tended to favour manual gearboxes (Stick shift), whereas the Americans and Canadians tended to favour automatics.

I do find it quite funny that although a diesel powered engine is more economical and the favour of the environmentalists for the best part of the last ten years, all of a sudden it has fallen out of favour, and we are being told to sell them.

Right, so let's get into this new terminology. We have 'conventional' cars, HEV, PHEV and EV.

HEV stands for Hybrid Electric Vehicle. How this works under the Bonnet (Hood), I'll come onto in a bit, but all you need to know for now is that it is powered by both a combustion engine and an electric motor. It, therefore, has a fuel tank and a battery. The battery is charged by regenerative braking only. I'll cover that in more detail later.

PHEV stands for Plug-in Hybrid Electric Vehicle. The only difference here is that you can charge up the battery before you leave home.

EV stands for Electric Vehicle. This is powered solely by an electric motor and has a battery to store the energy. Some EV's do have a small engine called a range extender. These are generally motorbike size engines that have small fuel tanks. This helps to relieve peoples 'Range anxiety'. 'Range anxiety' is caused by the fact that there aren't electric charging points every two miles like there are petrol stations. Also if you run out of battery, you can't grab a can out of the boot and walk to the garage to top up your car. You need to be plugged into somewhere that can charge you up.

Driving pure Electric Vehicles requires an attitude change. I want to cover that in more depth later so, for now, let's continue looking at all of the new technology.

Before I go any further, there is something I do need to say. From a driver's perspective, it doesn't matter what type of system you have or what combinations of the following configurations control everything under the bonnet. For you, driving and controls are just the same. I would say that there are only two main differences. Firstly, all of these cars are similar to drive as a conventional automatic. Place into drive, foot on the brake, release the hand brake, touch the throttle and away you go. The only other thing that you will notice is how quiet they are. I know the Nissan Leaf has a sound generator on the front so that pedestrians can hear it coming.

From inside, as you set off, there is no sound. It's like releasing the brake and allowing a conventional car to roll down a hill without the engine running.

So, how did all of this new technology first start? I worked for Jaguar Land Rover a few years ago on their Range Rover cars. I remember the first HEV cars being designed and finding out that their batter had a three-mile range. 'How stupid,' I thought, 'why put in such a small battery?' As it turns out, it was a brilliant idea.

I want you to picture in your mind a road. We will keep on coming back to this particular road for other examples. The road is one mile long and has three sets of traffic lights that turn to red as we approach them.

The first car that we use is ten years old. It accelerates away from the start line using fuel. It then uses its brakes to slow it down for the first set of lights. This turns the car's Kinetic energy into heat as the brakes use friction between the disks and the pads. It sits there with the engine idling for thirty seconds until the lights change and it sets off using more fuel to power itself forwards. It does the same process until it reaches the end of the mile where it has used a quantity of fuel.

Now let's use a modern conventional car with stop/start technology. This accelerates and decelerates exactly the same but, for each thirty-second stop, it used less fuel as the engine is turned off. This car would have used a minute and a half of idling fuel less than the car above.

The HEV car starts off with a battery empty. It powers away using the engine but, to slow down for the first set of lights, it uses its regeneration ability until it is down to a few mph when the brakes take over. The engine would have turned itself off the moment the driver took their foot off the accelerator. As the lights change the car accelerates but, having some energy stored in the battery, the car would accelerate to say 25 mph before the battery ran out and the engine takes over. Cars use most of their energy accelerating, especially from standing still. Range Rovers, and other large cars, can weigh up to 2.5 tons. Can you imagine the amount of energy that it takes to accelerate one of these to 30 mph? Fortunately, it doesn't matter what the car weighs as you are turning the same amount of kinetic energy (motion) into electricity, then using that electricity to accelerate the car. There are inevitable losses due to minor inefficiencies but being able to re-coup 25 mph from an initial 30 mph is really good. You would only be using fuel to accelerate you the last 5 mph.

Usually, the battery won't be empty whenever you get into the car so even first thing in a morning you would be able to get some acceleration from what is in the battery. These cars generally have the smallest batteries. Even the ability to drive two or three miles on pure battery power will make a huge difference to your fuel costs. To put this into rough numbers if you had a car that would typically give you 40 mpg around town, this could increase to say 60 mpg. Obviously, if you did more motorway driving, then you wouldn't see such a big difference.

The PHEV cars are the same in operation except that they generally have bigger batteries. These can give you a range of 30 to 50 miles meaning that if you have a relatively short commute, you could actually go to and from work, the shops etc. on purely electric power and not use any fuel. The Mitsubishi Outlander is one such car, the manufacturers claim that it can return up to 148 mpg. Now that is amazing.

Before I get onto pure Electric Vehicles, I want to give you more of an insight as to how both HEV and PHEV cars work. Some cars have a conventional engine set up driving the front wheels and electric motors on the back wheels. Others, again the Outlander is a great example, is basically an electric car with a petrol powered generator. The car constantly runs off the battery, but as the battery runs out of energy, the engine starts up and tops up the battery. This can be used in several ways. In the next couple of years, London is banning fuel powered Taxis. The engine can be run in the Outlander to charge up the battery fully. As the driver enters the exclusion zone, he can turn to purely electric power and have a clean car. Plugging the car in overnight tops up the battery for the following day. So this car gives you the short range (50 miles ish) of pure electrical power, then a conventional tank worth of fuel to enable you to complete a full day driving without refuelling. This would actually be beyond most people's bladder range.

One of the features that I particularly love about the car is when the engine turns on and off. Let's say you need to accelerate onto a motorway uphill and you need to achieve motorway speeds. As you start to accelerate, the control system will realise that it is unable to provide you with the power that you need on just the batteries, so it starts up the engine. This gives you the boost that you need. As you ease off the throttle (assuming that there is enough battery left), the engine will stop. It can literally stop and start whenever the extra boost is needed.

This brings me on to the engines that can be used in these types of cars. Conventional cars require an engine that will tick over slowly but can also rev to a high enough limit so that you can accelerate sufficiently between gears so that you can manage with only a few gear changes. Electric PHEV cars with electric generators only require an engine to start and run at one speed. Regular car engines are designed to take all sorts of speeds and loads. A generator will start up and accelerate to 3,000 rpm and sit there. The engine won't be overstrained doing that. In fact, it can be designed to be a lot lighter as it doesn't have to be able to take the strain of high revving like an ordinary engine. Actually, it opens up all sorts of possibilities for designs. Standard petrol (gasoline) engines are only about 30%

efficient. Let me put this in a way that you will be able to visualise. Picture in your mind a typical four-cylinder engine. The amount of power that it provides to the gearbox is equal to the power output of one cylinder!

Petrol and diesel are very energy dense fluids. Each gallon of petrol has the same energetic (explosive) potential as seven sticks of dynamite. A typical UK car will carry 10 UK gallons, around 45 litres. That is the equivalent of 70 sticks of dynamite. That is a lot of explosive force to carry around with you. Although it contains a lot of energy, we waste a lot of it in producing heat that we throw away, and as noise, but most of it is used to turn the engine itself.

Just before I go on to the advantages and I will be honest, disadvantages, of Electric cars I would like to take you on the journey that a litre of petrol takes. The crude oil is refined in one of the massive oil refineries that consume so much energy, it is estimated, that each gallon of petrol has already wasted 6 kWh of electricity by the time it is made. It has also given off the equivalent of 500 g/km of CO_2 before it even gets anywhere near an engine.

So what is the difference between how fuel and renewable energy is produced?

How to make petrol (or diesel)

Search for it underground.
Use a big drilling rig to drill a large borehole.
Pipe it onto a ship.
Travel across the ocean(s).
Pump into a refinery.
Refine.
Pump into a tanker.
Transport to the local garage.
Pump into a car.

Negative effects

Personal cost
Carbon dioxide
Other harmful chemicals

How to make electricity (for your car or home)

The wind blows
The sun shines

Negative effects

None

I'm not sure if the 6 kWh figure includes all of the transportation or is just from the refining process itself.

To charge up your car on a sunny day you plug it into the car charger on the front of your house and have the solar panels turn sunshine into driving energy for you.

Yes, the solar panels consume energy when they are made, and I don't know how much but I top up my car every week. It costs me £112 a week. At £1.32 per litre that is 85 litres or 19 Gallons. Assuming I work 48 weeks a year the fuel that I used would have consumed 5,472 kWh of energy to make it, so there are the greenhouse gas issues to consider, and that is before I burn the 1,615 gallons of petrol that I will burn in my car. Therefore, costing me a total of £2,132 a year.

So let's look at an all-electric car. Let's begin with a fuel cost comparison with the figures above. I get about 40 miles per gallon out of my car. That means that the £112 gets me about 760 miles a week. The New Renault Zoe has a 41 kWh battery and has a reported range of 250 miles. Let's say, in the real world where I drive my car, it is 180. My car is supposed to do 55mpg, but I get 42.5mpg, so I'm using the same ratio. 180

miles divided by 41 kWh equals 4.4 miles per kWh. The worst price that I can find on the internet today is 17 pence per kWh. So to charge this Zoe from empty to full would cost me £6.97. The Zoe only has half the range of my car, so £56 divided by 2 is £28. So petrol is four times more expensive to use than electricity when compared to the highest price I could find. Night time economy 7 prices can be as low as 9 pence per kWh. That would then cost £3.70 making my petrol cost 7.5 times as much.

Don't forget, this is before we start looking at the environmental impact. Let's assume that I use the high rating to charge my car and I drive it 260 miles a week. This is quite shocking really. It would cost me £10.07 at the worse rate and £5.32 at best. Basically, that would save me nearly £1,000 a year. Actually, it would be more as I would charge my car at home from my solar panels so I would be topping it off for free, then a half charge during the week so that I had enough to finish off going to work for the week, and enough electrons to get me home. So one charge of, say 20 kWh, would cost me £2.40. I could go to work, and I work away from home during the week, for £2.40 a week and fill from my solar panels, instead of paying £34 a week as I do now. £115 a year to get to work instead of £1,632, now that is a nice saving. Let me put it this way, in my current situation that would mean that for the same price as I pay now for my transport, I would have £126 a month towards my electric car.

Actually, I now drive over 700 miles a week and currently spend over £400 a week on Petrol. Changing to a brand new Zoe after selling my current car that I pay £200 a month on, I would save a fortune and be making a big difference to my impact on the environment.

Before I get onto the other advantages of having a fully electric car, let me cover the disadvantages. When the Nissan Leaf came out around 2013, it had a range of 80 miles. If, like me, you don't like emptying the tank on your petrol car then you would probably want to top it up when it gets down to a quarter full. This is fine if you have a reasonably short commute and can charge the car up at home. If not, then you are looking to charge it up every 60 miles. Actually, it gets worse than that. If you are charging en-route, then there is another problem. Unlike topping up with petrol, electricity doesn't run in at a constant rate. Think of it more like pumping up your tyres. If your tyre is flat, a little bit of air will make it look pumped up. But it takes a lot more to bring it up to the right pressure. Fast charges will charge a car up to 80% within 20 to 30 minutes, depending upon battery size. So now our 60 miles of the useful range has dropped to 40 miles between stops. This is where range anxiety comes from. Can I get to where I can charge the car up before I run out of electricity?

This is where a whole host of new, and upgraded cars, are game changers. The original Renault Zoe started off with a 20 kWh battery. This was a useful commuting car but would be limited as per above. The new Zoe has a 40 kWh battery with a claimed range of up to 250 miles. Even taking into account typical driving styles and undulating terrain, it is proposed that this would give you a real-world range of around 180 miles. Let's assume that you still want to top up at a quarter full, that gives you 135 miles until your first charge. That's over 2 hours driving at motorway speeds. Then an 80% charge and stopping at a quarter of a tank gives you 100 miles between further stops.

Let me put that into a real-world scenario. When I drove 12,500 miles all over America in 2012, I was driving 300 miles a day. It is possible to drive a lot more miles in a day, but I was doing this day after day after day, so I had to limit myself to what I could continuously manage. Given the new Zoe's capabilities, I would drive nearly two hours then stop for lunch. Drive for another hour and stop for a coffee and a comfort break. Then drive for another hour to arrive at my destination. That would be 335 miles.

A car with a 250-mile range is as capable as any other 'conventional' car.

Here is something that I think is super cool. If you bought the original Zoe with its 22 kWh battery, you can replace it with new 40 kWh battery. In the last few years, the battery technology has moved on so much that the energy density has increased to the point where the new 40 kWh battery is the same size, and only slightly heavier than the original one. What do they do with the old battery packs? I'll come on to that later on in the book.

Let's have a science lesson. So what is a kWh? The 'k' stands for kilo, as in a thousand, 'W' stands for Watt and 'h' is for an hour. A Watt is, according to what I found on Google, one Volt of one Amp for one second. Ok, so I'm a mechanical engineer, and I don't understand that. Let me put it another way. Houses used to be full of filament bulbs, these were the standard house type of bulbs that we all used to use until about ten years ago. A standard room bulb was a 60 watt. But if you wanted a lot more light, you would use a 100-watt bulb. Now let's imagine that you have ten of these 100-watt bulbs, it would equal 1,000 watts, or a kW. If we have these ten lights on for an hour, they will have consumed 1 kWh of electricity.

In the UK we are used to quoting cars as MPG, miles per gallon, anything over 40 is classed as good. The higher the number, the better. In Canada, being metric, they use l/100km, litres per 100 km. My first car in Canada was a Ford Windstar, think Ford Galaxy or Vauxhall Zafira. In this, I used to average 11 l/100 km. I then bought a Nissan Murano and this used only 5 l/100 km. In this case, you are looking for a smaller number. I'm not sure what the standard measure will be for EV's. As I show above, the Zoe does 4.5 miles/kWh (5.95 miles per kWh based on the official range of 250 miles). That gives you an idea of what sort of numbers to look for economy wise.

The other way that we look at cars is by the amount of power that they produce. I'm going to apologise here, but it's going to be another science lesson. BHP or Brake Horse Power does actually go back to the days of horses and calculating how much they can lift vertically while pulling forwards with a rope attached to a harness. I won't go into any more details for now, but we use it to measure how much power an engine produces. It is actually only half the story. If I put a 1,000cc motorbike engine that produces 100 hp (the shortened version of bhp), into a car that had a 1,000cc engine of only 60 hp, it would struggle to power it. Why, because it doesn't have enough torque.

So what is torque? For me, it is harder to explain than to give you examples. Petrol engines produce their highest torque at higher revs, which is why they are generally used for racing cars. Diesel engines produce their highest torque at lower revs, which is why they are used for lorries and buses. They are great at helping them set off and climbing up hills. A diesel engine with a turbo gives you the best of both worlds. Let me put it another way. In a diesel car, you will change gear less because it has more torque power at lower revs, i.e. normal motorway speeds. As you start to drive uphill, you ease down on the accelerator more, and you are easily able to maintain speed a lot easier. Whereas, in a petrol car you might find that you have to change down a gear to maintain the same speed.

So how can we equate ICE engines to EV's? Let's have a look on the world wide web and see what we find. I've pulled up the stats on the 2017 1.0 l Nissan Micra. I've owned a couple of Micra's over the years, I have loved them. So this one is a 1,000cc, three-cylinder engine producing:-

72 bhp and 70 lb-ft that is 54kW.

Let's have a look at a second car. My parents recently bought a second hand Mercedes C180. Some stats I've just found for a 2.0 l petrol read like this:-

142 bhp, 155 lb-ft that is 106 kW.

Electric cars are a lot different to drive, with respect to power. Electric motors are all about the torque, so never try to race an EV off the line. He will win, every time. Plus they will be able to generate the extra speed into spare miles, and you will just get to heat up your brakes.

Let's look at the Nissan Leaf, the car that really started to revolutionise the UK.

80 kW (110 hp), 280 n-m (210 ft-lb).

I've just finished typing that into my pc, and I've looked at the figures, WOW. Here is a car, that is a bit bigger than the Micra, but a little shorter than the C180. To be fair though, the interior is probably not a lot different. The Leaf has 50% more hp than the Micra and about the same as the C180 but look at the torque. It has three times what the Micra has and 25% more than the C180. I will be honest with you here. I love all-electric cars, but I didn't know what they would be like towing a caravan. Obviously, it would burn through the battery really quickly at motorway speed, but it has more than enough torque. I'm imagining what we could have in five years. A similar motor but a battery with a lot more kWh and away you go. Plus the added benefit of being able to charge up at the campsite.

There are actually caravans being produced that have batteries and motors. These basically power themselves therefore not affecting the range of the towing car.

To be honest, one of my dreams is to have a mid-size motorhome that is covered in solar panels that, on reasonable journeys, can be fully self-sufficient power wise. Imagine driving to the continent, topping up as you go but touring say, 50 miles a day and all of that being off the solar panels, including your lighting, hot water and cooking. Oh, and maybe throwing in your air conditioning.

I digress; let's get back to looking at cars.

Let's now look at some of the advantages of owning an EV car. How many of these things do you recognise? Spark plugs, oil change, oil leak, air filters, servicing, radiator leak, water pump leak, head gasket, brake pads, brake disks, worn big end, emissions, fuel leak, exhaust broke and who remembers the petrol tanker strike from the turn of the century? With all of those things, have you ever had to pay out good money to get them repaired? I know I have, each and every one. Well on a pure EV, apart from getting your car serviced, and we will come back to that in a minute, none of those can go wrong so you won't have to pay for them. This is another area where green cars save you green.

It's funny, I was just watching a car repair program where they took older cars, repaired them and spruced them up in the hope that they can sell them for a profit. They had a couple of cars in that needed repairs to the gearbox, the top of the engine and the brakes. Yes, electric cars do still have brakes, but when you think that around 90% of the energy absorption is done by the regeneration system, then I would think it is possible that the brakes and discs could last the life of the car. These repairs to the cars cost hundreds of pounds and, apart from the brakes, are not on EV's.

I'd like to introduce you to the Tesla X. This car is fantastic and is my dream car. It is a four-door saloon. If you open the doors, it will look like a pretty much standard five-seat family car.

Open the boot (trunk), and you will see a nice large boot with an extra deep section, and when we look under the bonnet (hood) we will find nothing! Tesla calls this space the Frunk, meaning it is the front trunk. Then your brain goes into overdrive, where are the motor and the batteries and all of the electronic control boxes. That is all hidden out of the way. The battery is in a long, wide, shallow section box that is under the passenger cabin. The motor(s), it can have one or several motors, are located in the boot. Do you remember the extra deep section of the boot, well the part in front of that is the motor? In front of the motor is a reduction gearbox as motors have higher rotational speeds than the wheels, then there is a differential. That's it, there is no need for anything else engine related. Even the reduction gearbox is only a small gear turning a big one.

As part of my research for this book, I took a trip to my local Tesla car showroom. There are currently seventeen showrooms, according to the Tesla website, in the UK. Eighteen Tesla stores are open. Just while we are on statistics, there are, according to Zap-Map, just over 6,600 charging locations in the UK, this is an increase of about 745 in the last 30 days, 1st November 2018. But, the number of car charges at those locations is 18,657. That's an increase of 25% in just under a year.

Compare the stats above to the figures in the early part of 2017. Zap-Map gave the following stats on 2nd April 2017; 12,330 connectors, 6,652 devices at 4,355 locations in the last 30 days, that has increased by 196.

Tesla offers an awesome business opportunity. Let say you have a hotel or are part of a collection of small business in an idyllic location. If they approve you, Tesla will come and install two Destination Chargers at your location for free! This will, therefore, attract Tesla drivers to your location. Using the charger, they would be there for a minimum of twenty to thirty minutes topping their car up. With the current prices of Tesla's, these people are likely to have, how can I put this, more disposable income than some other people. I think it would be an awesome addition that would increase footfall to a location. Anyway, let's get back to my trip to Tesla.

I was met with great enthusiasm by two great guys who, unfortunately, I can't remember the names of. There were several things that we discussed quickly. I was interested in asking if the showroom was part of Tesla UK or a Franchise. I know franchises can be very successful in other types of businesses, including the motor industry, but I was quite surprised to learn that Elon Musk's philosophy was to keep all of the Tesla family together. What this means to the consumer is that it doesn't matter which showroom you visit, their cars, and other products will always be the same price. The other thing that I was impressed with is that the guys I was talking to are not car salesmen, they are Product Specialists. With all of the research that I had done for this book so far, I thought that I would be just asking them a couple of questions and confirming what I currently knew. Oh, how wrong I was. While I am not going to go into too many details of each car, I will give you an insight into what you can expect when you own one.

The first question people ask is about the range of cars. In 2013 there were only about 3,000 car charges in the UK and the cars at the time had limited ranges of only up to 80 miles. That is where range anxiety has come from. Let's dispel that myth right now. As I pointed out above, there are now many more thousands of chargers, I would say that most, if not all, town centres would have at least one set of charging points. While London is scattered with charging points, you might be surprised to find that out of the way places like Ulverston, Penrith and Carlisle in Cumbria, Bowness in the Lake District and all of the car showrooms that sell EVs, have a couple of charging points that you can use as well as the supermarkets that are installing them. Asda is adding them to all of their new stores. I know that Morrison's near where I work has them and that Tesco's are installing them. Large supermarket organisations are very business savvy. If it brings them more customers who will spend money in their store, then they will add them.

So what sort of range have Tesla's got? The New European Driving Cycle, NEDC, is a set of tests that all of the manufacturers use to conduct their tests. While the results might not reflect real-world driving, they do at least give you the opportunity to compare all cars against one another. It's like the fact that my car is supposed to be able to do 55 mpg according to the brochure. But we all know that in reality, the car will do less. I know in the '80s all cars fuel economy figures were quoted as 56 mph. So if I drive on the motorway at 70 mph, I will get less mpg. If the motorway goes uphill, I will get quite a few less.

The new Renault Zoe that I've commented on earlier has an NEDC range of 250 miles, but Renault states that you can expect 186 miles. The NEDC range of the Tesla Model S P100D is 380 miles. The guys that I was talking to stated that this is probably more like just over 300 miles, I don't know about you, but I couldn't drive for nearly five hours without stopping.

Another thing that people are concerned about is when and where they go to fill up their electric car, because they often run out of petrol. I could comment that an electric car is perhaps not for them. But let's look at their typical week. They go and do their weekly shopping on a Saturday at the local supermarket and top up the fuel tank while they are there. Monday they drive 50 miles to and from work, the same for Tuesday, Wednesday and Thursday. Friday morning they set off, the red light comes on, they haven't got time to stop, and they end up running out of fuel before they get to work. Let's look at a different scenario. They set off for work Monday in a nice warm car that is showing 99% charged. Tuesday it is 99%

charged and so on throughout the week. A lot of companies are now offering to charge at work too.

I'll tell you another little story that I hope they won't mind me sharing. Nissan has a van/people carrier version of the Leaf. It is called the NV200. You can buy it with petrol or diesel, but the electric one with the Leaf electronics and battery looks amazing. British Gas has apparently replaced 10% of its fleet with these (2017). It means that they are fully charged and ready to go first thing in the morning. Well, one guy was so impressed with his van that he decided to get himself a Tesla. Now that is what it is like being so impressed with a new type of product.

Right, let me ask you another question. Have you ever had to treat one of your cars for rust? Have you ever had to fill in a bit of the bodywork, or have you ever had parts fall off? I remember back in the '70s my dad wanted to check the brakes on our car. He went to jack up that corner of the car, he positioned the jack and started pumping away. After a while, he realised that the car wasn't lifting. He stopped and went to open the door to see what had happened to the top of the jack. The door wouldn't open. The jack head had pushed its way through the jacking pad, through the sill and in through the bottom of the door. He bought a newer car that week and scrapped the old one. I don't know if it was the same one, but we had one that he could only put 4 gallons in as there was a hole in the tank!!

Cars rust because they are made of steel. Other metals corrode too but are not as quick to rust like steel. All Tesla's are made out of aluminium which doesn't corrode anywhere near as aggressively. The Tesla 3 that is coming out soon does have some steel components to save cost, but the body, which is the thinnest material, is still all aluminium. To save weight a lot of aircraft leave most of their aircraft bare of paint so what you are actually seeing is the aluminium alloy that the aircraft is constructed of. If a car is made out of aluminium, then it is going to last a long time. I've seen rusty patches on cars that are only 4 to 5 years old. It is possible that these aluminium cars could last ten to twenty years, as long as the paint holds up.

Battery life is something that people always question. The engine in a typical family car would be lucky to make it 150,000 miles. Some last a lot less and the odd one goes on for a lot longer, but I'd say that 150,000 miles are a typical average. After this, you are looking at an engine that has worn pistons, piston rings and barrels. It would need a lot of work to bring it back into its initial power range. Tesla guarantee that after ten years of average driving their batteries will have a minimum of 70% of their original capacity. This is an ultra-conservative figure because these batteries, and indeed all of this technology, is still so new. There is a car in America that is a 2015 model that has done 200,000 miles, and his batteries are showing only a 6% decrease. If we do an average of 12,000 miles a year, then that would take us nearly 17 years. That is less than 0.5% loss every year.

With having regenerative braking, I knew that the brakes wouldn't be used a lot, but the current projections show that they are expected to be replaced every 80,000 miles. It is possible, with a lot of motorway driving to achieve 60,000 miles out of your brakes, according to the Kwik Fit website. But it is also possible to only get 25,000 miles with lots of stop-start driving and high-speed driving.

So what about warranties? As I was informed yesterday Tesla currently gives an 8-year unlimited mileage warranty on the motor and batteries, and 5-year 50,000-mile warranty on the rest of the car if I am reading my notes correctly.

Let's look at another exciting area of technology and what it can really do for us. A number of years ago I slipped on a wet floor and hurt my wrist as I tried to stop myself from falling. I drove myself to the hospital, and the X-Ray confirmed that I had broken my wrist. The next thing the medics did confused me. They asked me how I got there. I told them that I had driven there. They then informed me that I couldn't drive with my arm in a cast as I wasn't insured. So I'm a single guy, my children are too young to drive and didn't live nearby anyway. My parents lived an hour and a half away, and there was no-one else I knew to help me get home. I needed to get home and somehow have someone drive my car home. What can I do, I'm stuck...

Let's look at that scenario with any car that has level two auto drive. As I was still physically able to monitor the car, it would be able to drive me home legally.

How about a different scenario. My parents have driven cars all of their adult lives. In a few years, their reactions and eyesight will get to the point where they will no longer be capable of driving themselves around. Not only will this cause them difficulties getting to the shops, but it will also cause them difficulties when they just want to go out. With auto drive level five, they will still be able to retain their freedom. They might need to get the car off the drive safely and have to park it at the supermarket, but all of the on-road driving will be taken care of by the car. Even when driving around the car park, the car will be in full control of preventing accidents to vehicles and pedestrians.

There is a fantastic video on 'fully charged' about self-driving cars. Nissan brought a prototype self-driving Leaf to the UK and took journalists out in it to prove how good the technology is. The difference between the Nissan system and the one in the Tesla, is that the Tesla one maps all of the miles that it covers and transmits that information to the central computer. All of this information is then transmitted to all of the cars so that, even though you have never driven down this road before, your car will avoid the pothole. The current Nissan system will put you through it every time. As of April 2017, Tesla is recording 10 million miles of roads per day!

So what next for Tesla cars? As I write this book, the price of the new model 3 has been announced, but we are not likely to see one in the UK for a while. Over 400,000 orders were placed in the first week that the cars were available to order. But, being such a minority in their vast order book, we will have to wait until they do the production run of the right-handed cars.

Chapter 3, Electric Cars, Part Two

In this next section of the book, I am going to start off by telling you about the test drive that I had in a model X. If you are seriously thinking about getting a Tesla, go and talk to them. I'm sure that they will be willing to let you take one out for a spin. I, therefore, feel very honoured to be offered a test drive, as I am just a lowly author.

Awesome doesn't begin to describe the experience that I have just had. Three weeks after my first chat with the Tesla team in the Knutsford showroom, I was back again with my daughter, Amy, in tow.

I must apologise for everyone that knows me, because all that I have talked about for these last few weeks, is driving the Tesla. I'd shown Amy numerous videos, but I could tell that she still didn't get it.

Let me go through some details before we get to the driving. The model that we drove was the 'X', and it was a P90D, Let me explain what that means. The model X is a four wheel drive car and therefore has dual motors, and that is what the 'D' stands for. There are a set of two that drive the rear axle and a single one that drives the front. The 'P' stands for performance model, which therefore have the Insane and Ludicrous modes. I might be slightly wrong there. He did mention something about one of the cars having a line under the 'P'. That might be the one with Insane mode. If you want the full details on that have a look online or pop along to your local Tesla dealership. They will be happy to help you out. Anyway, the only other part of the numbering system that I have not covered is the '90.' This means that it has a 90 kWh battery. This means two things; the amount of energy that it stores but also the rate at which it can discharge it, therefore, your ultimate acceleration. The Model S with P 100 D can accelerate from 0 to 60 mph in 2.5 seconds. I think that I have got that right. There is an update to that, the Super Insane mode, or whatever it is called, drops that to 2.4 seconds. But, as the warning goes, this can damage your motor and battery!

For me, anything under 5 seconds is more than good enough. My Dacia Stepway, that weighs just under a ton, does the 0 to 60 in about 12 seconds. The three-ton Model X with three of us in was super rapid. It's funny I was laughing at one of Robert Llewellyn's 'Fully Charged' videos yesterday about how the car that he was driving could accelerate so fast that it took his breath away and his vision went. Well, this car that I was driving today was not as fast as the one that he had borrowed, but it still went from 20 mph to 'oh my gaudy' in about half a second. I certainly accelerated at a rate that I had never experienced before, but also at a rate that would typically only be found in a quarter of million-pound supercars.

I started off in the passenger seat as we were driving out of the quite tight car park, and as we drove, some of the controls were explained to me. Having driven various cars in America, I am fairly used to different gear lever configurations. Whereas the standard UK car will have a gear lever between the two front seats, regardless of whether it is a manual or automatic, a lot of American cars have a gear lever behind the steering wheel, effectively in the right-hand stalk position. Although there are about five positions for the gear lever, there are effectively only two gears on this car, 'D' Drive and 'R' Reverse. Because an electric motor isn't constantly running, there is technically no requirement for an 'N' Neutral. There were some other letters, but we don't need to concern ourselves with them for now.

So, seat belt on, heart rate elevated. The only other trip that I have had in an electric car, was a whiz around a car park in the passenger seat of a Nissan Leaf.

We cark in a quiet car park of a garden centre and change seats. Now I am in control and confused. The car is already on, apparently! Flick the gear lever down until it is in the 'D', take your foot off the brake and nothing happens. The parking brake is electronic, so it's a slight touch to the accelerator, and it's off, and the car starts to roll forwards. A few years ago I had a 7 series BMW. It had a 4.8 l, V8 engine. It had so much power that it was awesome. The first time that I drove it I was nervous that it would only take a slight touch of the throttle and it would be lurching away from me. I was pleasantly surprised that it was nice and gentle and smooth. Tesla's are just the same. You drive it like any other car, add a bit of throttle more or less to help get you to your desired speed at the acceleration that you want. It is when you push the throttle lever further or fully down that the world starts to get all fuzzy in these cars.

We set off out of the car park nice and easy, you wouldn't have guessed from the outside that this car was any different to any others that are on the road.

I got us nice and settled on this road, plenty of road ahead and nothing behind us. This was when I pushed the throttle hard. The only other time that I have ever experienced G acceleration like that, was when I was strapped into an ejection seat in a Jet Fighter aircraft on the start of a runway, in my RAF days.

Phenomenal doesn't begin to describe it. Have you ever been stuck behind a tractor or something else on the road and known that you have not got the power to get past it? Well, you wouldn't have that issue in this car. Oh, and we were in ordinary mode!

Tesla's are fantastic cars. There is no denying that. Are they perfect? No. But then again, they are in their infancy. Electric cars of this sophistication have only been around for a few years. Every day they are fixing minor issues and making gradual improvements. I remember when they first started putting diesel engines into cars, they were a bit clunky but, after a few years further development, they became extremely efficient, very smooth and incredibly powerful. But I reckon that, in total, it took them ten years to achieve that level of technology. I just wonder where the electric car market could be ten years after they start.

The last thing that I want to mention about my test drive, was the automated driving. I give a talk on automated driving that is available to U3A groups throughout the country, and anyone else who would like to hear me talk. Contact me for details. I explain to them why everyone in their senior years, who wants to maintain their transport freedom, should have a car with self-driving capabilities. Not only to help them get about but also, and this actually applies to people of all ages, to help protect those around them.

When you reverse out of a parking space in a supermarket car park, you have to swivel your head fully left, and full right, to enable you to see all of the pedestrians, cyclists and other cars, that might be near you. With a self-drive car, you will still need to maintain your lookout, but the car will also be looking in all directions, at all times, for you. It makes you a safer driver to be around.

On our way back to the showroom we encountered two situations. The first one was a red light at some road works. The second one was a queue before getting to a roundabout.

I had the car set in auto drive at 30 mph. The lights at the road works were on red, and there was a car already waiting. Sensing the car far ahead, the Tesla slowed down, coming to a stop a nice safe distance behind the parked car. I figured that I would have to initiate something once the traffic lights changed to get the car to set off. I was wrong. It accelerated and followed the car in front all of the way to the queue at the roundabout, maintaining a good location within the lane. As the car ahead slowed down, so did we until we came to a stop. As the car ahead nudged forwards, so did we.

This car takes away the frustration and monotony from you and instead wraps you in a blanket of protection.

I could go on about the minor details that I had on this drive today, but really, it's up to you to go and experience it yourself.

I really must get myself one. The smoothness of the ride and the power delivery and recovery coupled with the level of electronic complexity is way beyond any systems that I have encountered, including some of the more modern highly sophisticated aeroplanes that I have flown. All of that added together in a car that will last a long, long time and is as good as we can make it today, with respect to environmental impacts, in its production and everyday usage. It is, I believe, the standard to which all manufacturers should aspire to emulate.

I would like to thank Tesla UK, and the team in Knutsford, for the opportunity to drive this amazing car, and for their time in helping me to understand some of their technology, so that I could pass it on to you, the reader.

After the test drive, I asked my daughter if she had any comments for the book from her position sitting in the middle seats. The first comment came very quickly to her mind, "When are you ordering one?" Her second comments were about the comfort and feel of the car from there. She said that there is loads of room. But one of the things that she really noticed was the ride. As I mentioned before, this car is on air suspension, so you don't really feel the irregularities in the road.

All electric cars are automatics. Well, that is what they are like to drive. They are actually very different from automatic cars mechanically. The reduction gear on the Nissan Leaf is effectively 8 to 1. As the car is stood still the electric motor is directly connected to the drive system continuously. There is no need to have a clutch as both a manual and an automatic car would. An engine at idle needs to turn at around 600 to 700 rpm to keep on going, an electric motor doesn't. When you set off in a conventional car, there is limited power to get you started, as the engine doesn't produce maximum power until it is in the 3,000 to 4,000 rpm range. Electric motors, on the other hand, have high torque from the moment they start moving. You will find this out if someone sets off next to you from say, traffic lights. It is also a useful feature if you are going slowly uphill. You also can't, theoretically, stall a motor, certainly not like an engine. I say theoretically, but if you were pulling a heavy load, like a caravan and went up a steeper and steeper road, the motor would eventually stop turning and stall. For normal roads and loads, no, you can't stall electric cars.

My current car, that I bought recently, has stop/start. I like it, I don't waste fuel sitting at traffic lights, and there are a few on my commute on the way home from work. Last week I was a bit too quick for it, and I managed to stall it.

Why, you might ask, if I am so big on electric cars and why didn't I buy one instead of a petrol one. There are two reasons really. Firstly, my weekly commute involves driving 160 miles a day. The second-hand cars that are coming onto the market don't have the range for me to do this without stopping and I couldn't afford to buy a new one. Well, I couldn't get finance on one that had sufficient range for my current lifestyle.

My second car is a different story. I currently have a Smart car, (you can love them or hate them, I don't care). My daughter, who is 25, is just about to pass her driving test, will drive the Smart car for a couple of months, then I will be replacing it with either a Leaf, or a Zoe. These can be found in the UK for around £5,000.

Update, as I write at the end of 2018, she has owned a Nissan Leaf for over a year. She loves it. There are a couple of issues due to where I live, but I will cover those later in the full update section.

There is something that needs to be taken into account, and so far, I don't think that websites like Autotrader have added this to their search options. When you buy an electric car, you can either buy or lease the battery. When you are buying a second-hand car, you need to be clear if the price is just for the car or if it is for everything. It does seem strange in a way. I'm buying a car but renting the engine. It's not how we are used to doing things. But, it is the way of the new world. On Robert Llewellyn's YouTube page 'Fully Charged' he suggests buying the battery (or leasing the battery and the car from one company). That way, if there are any problems, or you are involved in an accident, you claim through one source.

Nissan's second-hand car prices all include the battery. When my daughter took out her lease, the £130 per month covered everything.

Even if you buy the battery, most manufacturers guarantee that it will still have 70% capacity in ten years. You might think that number is high, or you might think it is low. What a lot of people don't realise is that although your car had 100 hp when it was new, and it states 100 hp in the owner's manual, if it is ten years old then you will be lucky if it is still producing 70 hp, and it won't be giving the same miles per gallon as it did when it was new. One of the differences here is that the electric motor should be giving the same power, it's just that you won't have quite the same range.

So let's look at what happens if you have a problem with your battery. How a particular battery is put together is up to the manufacturer but, as far as I am aware, they are all modular. This means that if one cell breaks down, firstly it will only affect your range slightly and secondly, that cell can be replaced. It's not like current car batteries where you have one large battery. If you have a problem with a cell in a lead acid battery, generally it means that you have to replace the entire battery.

Speaking of car batteries. There is one in electric cars too. The voltage of the driving battery is so high that it can't be used for utilities such as lighting, windscreen wipers, radio etc. So this is run off the car/leisure type battery. The Nissan Leaf has a nice option where you can have a spoiler fitted that has a solar panel. This is used to top up this battery.

Here is an often asked question, what happens to all of the old batteries. They are all fully useable either in power banks like those used to power Las Vegas or for power wall installations like you find in people's houses.

So what other technology are they putting into electric cars? Put simply, everything. There seems to be, at the moment, a vast technology increase in all cars. There are not many places around the world where you can have an automatic car starter to warm your car up in a morning but, because an electric cars heater is an independent heat source, you can warm your car on cold days before you get into it. Quite a few EV's come capable of being controlled by an app on your phone.

Consider this scenario. You wake up in the morning, it's dark, and you realise that there is a frost outside. You pick up your mobile phone and set the car to warm up. When you are ready to go to work, you go outside, you unplug the car from your wall socket and get in. The car is warm, your seat is warm, and in some cars, the steering wheel is also warm. If you want to minimise the energy usage from the leisure battery, you can turn off the car heater but leave on the seat and steering wheel heaters (unless you get chilly on the way to work). How good would that be to step into a preheated car every morning and night? I'm not sure what happens when the leisure battery is low. I presume it would have a trickle charge from the driving battery, but I don't suppose it's any different to the power requirements to run the air conditioning system in an ordinary car. With standard AC running you generally lose a couple of miles per gallon, so there is not much difference with an electric car, you will lose a couple of miles off your range.

Some pure EV's are based on current production cars. There is the Ford Focus and the VW Golf (the eGolf). Toyota and other companies have an EV in each of their range of cars. Then there are companies like Tesla, that are new start-ups, and only produce EV cars. But to be honest, it seems like the whole EV car phenomenon has taken the car market by surprise.

About ten years ago California brought out legislation to reduce the smog that cities like Los Angeles were suffering with. Car manufacturers had to offer EV's if they wanted to sell cars in California. This caused the manufacturers to produce something that ran on electricity. Toyota was one of the first manufacturers to look at this seriously, and they produced the Prius. The internal workings of the Prius are really remarkable. I wonder if a Toyota designer woke up one night shouting Eureka. This is because the car is neither an ICE (Internal Combustion Engine) nor electric motor powered. It is both. The gearing system allocates power from the motor or the engine depending upon the requirement of the driver and if there is any electrical power available. For the first few years, the car was only available as a pure HEV car without any plugin capability, but I think this has now been changed.

To be honest, for the first few years the industry and the public thought that the car was a joke. It was until Toyota sold its one-millionth car. That's when everyone started taking notice of it and the changes that were happening in the world.

It is said that if you want a car that is cheap to run, cheap to fix and reliable, see what taxi drivers are using. A lot of London taxi firms use Prius, and there are some now that are breaking the 1,000,000 miles mark. Why do they use them and why are they lasting so long? Because the engines don't take a lot of the strain, the motor does that so despite being high mileage, the engines are in good condition and don't cost a lot to maintain. Interesting side note, despite driving all of these miles, the motors don't require any servicing...

So as the Prius was becoming more popular, other manufacturers joined the HEV, PHEV and EV evolution. As I mentioned before, the Nissan Leaf and Renault Zoe came out around 2013. Honda has an EV, and there might be a couple of others, but I'm not sure who they are. Then the sales really took off. In the UK and most of Europe, EV charging points started popping up. What was really cool, was that these were free to use. So for the last few years, you could drive the length and breadth of the UK, and most of Europe, for free. Robert Llewellyn has an episode of 'Fully Charged' where he drove 1,000 km from Denmark, to his house in the UK, for five euro's. Unfortunately, we have now missed that boat, in this country at least.

How much does it cost to charge your car at one of these charging points? Firstly the industry is in its infancy, so everyone is trying different pricing options. Some are designed to charge so much for how long you are plugged in, some on the amounts of kWh you use. Some figures that I have come across indicate that you can charge from 25% to 80% in twenty to thirty minutes for about £6. While this is a lot more than it would cost charging your car at home, on something like the new Zoe that is an increase of 90 miles for the price of a gallon of petrol. Some places like hotels will allow you to use their chargers, if you have a meal or are staying overnight, for free.

Where can I find EV charging points? Firstly, they are popping up all over the place. I don't know if they are at all motorway services, but they are certainly at most. A lot of car dealerships that sell EV's have them, and these can be free to use. Because of the country's desire to reduce its carbon emissions, it is trying to 'electrify' the country, so even moderate size towns have charging points popping up. When you buy your EV, you will find one of the following two options. Either you can access via your phone a map of all of the electric charging points, or on your in-car GPS as in the Tesla X, you will see charging points en-route. It will also give you a display of what your charge will be when you arrive at your destination, so that if you start driving your car hard, the colour will change from green to amber indicating that at this pace, your charge will be getting low when you arrive. The display will also show you if the chargers are in working order.

Having mentioned Tesla, I want to spend some time looking at them and how they got started.

Elon Musk made a fortune from several companies, one of which was PayPal. Although he made a lot of money, he wasn't satisfied with just making money. He really wanted to make a difference to the world. As well as starting Tesla cars, he began Space-X with which he wants to send people to Mars.

Elon's goal with Tesla was the first to make a car that would make people sit up and take notice. Then he wanted to create an executive car that would take the world by storm. He certainly achieved that. Recently he completed his master plan, that is to produce a car for the people at a reasonable price. Let me try and describe what sort of following he has created with his cars. Without knowing the exact specifications of the car, on the day that people were able to pre-order the car for $1,000,

£1,000 or 1,000 Euro's. Thousands and thousands were ordered on the first day. The most amazing part about all of these cars is that they are all electric.

Tesla released the Roadster in 2008. This was an electric car that had a 200-mile range. It was very expensive at the time. Then in 2012, having refined a lot of what they put into their cars, they produced the Model S.

I mentioned to a lady the other day that I was writing a book about electric cars. "Oh, don't get one; my boyfriend has got one, and it's hard work," she said, and then went on to explain that she travels all over the country doing Mind Body and Soul events. Then she mentioned which car it was. I won't mention the actual make and model because I don't want to influence your decision-making process. I know that I wouldn't have got one if I was in their situation, certainly not for the sort of use that they are looking at. This brings up a very important point. Be very careful to select the car that best suits your needs. Let me give you a couple of examples. Last year, when my dad got his car serviced at the garage, they noticed that he had only driven 3,000 miles in the last 12 months. With that sort of annual mileage, and with the short trips that my parents go on, a Tesla S would not be the car for them. They would only need to charge it up once a month. I'd probably recommend either a Nissan Leaf or the shorter range Renault Zoe. If they travelled a bit further with the occasional long trip, then the BMW i3 might be better as it has a range extending engine.

Update, having seen me get excited about electric cars and coming on the Tesla test drive with me, my daughter decided that she wanted an all-electric car. We went to our local Nissan garage in our petrol Smart car. My first question was, 'we can see the sticker price of the car, but how much is the battery rental?' 'There isn't any,' came the salesman's reply. Nissan has a lot of first and second generation Leafs for sale, and they are making great offers. My Daughter has bought herself one for less per month then mine is costing me. The salesman told us that leasing batteries is something that Nissan and Renault tried to begin with but found that it really doesn't work. So all of the resale cars are sticker price only. Well, if you get a second generation car, they knock £1,000 off. Well, they did for my daughter.

Having now driven in it half a dozen times and being allowed to drive it, I don't know why I drive a petrol car. The Leaf is very high tech and as economical as advertised. Lastly, I nearly forgot, we are getting a free pod point fitted to the house too.

Just an update note, and something to remember if you are buying an electric car. They can't fit one if the cable is going to cross a public maintained path or road. In front of our house is a path that I do not own. Therefore my daughter has to plug in a cable to charge her car up. This works fine, but it is only charging at 3 kW. At this rate, it takes several hours to charge her car up. This isn't an issue as she charges it overnight. The house that she and her boyfriend are about to move into has a driveway so if she wants she can have a 7 kW charger installed on the outside of the house. This will mean that if she wants to give her car a quick charge during the day, it will charge up a lot more in a shorter time.

I see this no cable across paths being an issue in areas such as terraced streets. There need to be solutions thought of. This does though really depend upon how many miles you do a week. My daughter's new place is only a mile away from where she works instead of the current fifteen miles. She will now only need to charge up every couple of weeks or even once a month.

Right, let's get back to the original book. Her Leaf is good for my Daughter because she barely ever travels more than twenty miles each way, per day. For someone with a family who travels long range regularly, and can't afford a Tesla X, then the Mitsubishi Highlander PHEV might be the best option. Both the

BMW i3 and the Highlander are electric cars with petrol range extenders. If you wanted to buy a purely electric car, then have a good look around the market to see what is out there. I know that VW has an electric Golf and are bringing out a minivan that looks like a small VW Camper from the '70s. Volvo is also joining the EV revolution with several models. Smart car has an EV car in all of its Smart range as does Toyota. I'll do my best to give you as many options as possible, but new cars seem to be coming out weekly at the moment. I don't want you to feel that I am recommending anyone car or company in this book. I'll give you the information that I have, and you can go and make your own choice.

Something else that I want to cover is charging your car. To start off with, let's look at what you do with a fuel car. You put the nozzle in and hold it fully open allowing the fuel to fill your tank at full stream until it clicks off the first time. Then you pull the nozzle out a bit and pull the lever slightly to top off your tank. Therefore, the last couple of litres takes you longer to put in that the first couple of litres. Electricity is a bit different from a liquid. The picture in your mind, a multi-story car park, and a line of cars outside. The attendant opens up the gate, and the cars start driving in. The first few cars park where they like on the first floor. As that floor starts to fill up, some people take their time to park in between cars, and others drive to the next floor where there is lots of space. This goes on until the car park is 80% full. All cars now have to look for spaces, so the inflow of cars is a lot slower than it was when the car park first opened. When the car park gets down to its last few spaces, it takes a driver a long time to find an available space. Electrons are much the same. If you have a battery that is down to 25% and you plug into a rapid type of charger, and leave it for 20 minutes, you will find that you are quite quickly up to 80%. But, after that, the flow rate starts to reduce as the electrons struggle to find somewhere to park up. So when you hear figures of several hours to charge up, that is just to top off the battery. 20

minutes charging with a Tesla, for example, would put you back up to around 200 miles.

Here is a great tip that the guys at Tesla gave me. Mobile phone, tablets, laptops etc. are supposed to be good for 1,000 cycles. If you let your battery drop very low and charge it to 100%, that is one cycle. One of the secrets to fooling your device, or your car, is not to let it get too flat, and to not charge it to 100%. This will extend the life of your battery. With a Tesla, you have the option of stopping your charge when the car gets to a pre-set percentage. That means that setting your pre-set to 98% will prolong the life of your car battery. Phone shops and the like, will not tell you that as they want you to come back and replace your phone in the next couple of years.

There is another reason why you don't want to charge your Tesla to 100%. If the battery is full, there is no reason for the car to use any regenerative braking. Therefore you will be requiring more brake pedal to slow you down than you would normally by just easing off on the accelerator. Obviously, once you have covered a mile or two, the regen will work as you would expect. Speak to your local Tesla team about that.

The new road tax rules came out on the first of April 2017. For electric cars, these fall into two types. For those that are all electric and cost under £40,000, your road tax is free. If it cost more than £40,000 then it is free for the first year, then for the next five years, it's £310, after that it is free. This means that for the Nissan Leaf, Renault Zoe etc. the road tax for those cars bought after the first of April 2017, the road tax will be free until the next change of policy. The current Tesla range will be free of road tax for the first year, then £310 for the next five years, then free again after that. If you can afford to keep on buying new expensive electric cars, then you will keep on paying the road tax. If you buy a six-year-old car, then there will be no road tax to pay.

I think all of the above is correct; that is how I read the new regulations. Check before you buy your car. According to Google, Tesla has announced its new Model 3 price for the UK as £35,000. I thought this was going to be a lot higher. I think this price is set so that you can have some of the extras on your car and still be below the £40,000 threshold.

I briefly mentioned Zapmap earlier on in this chapter. If you are thinking about getting an electric car, then you and zap-map are going to become very good friends. Why, because it shows you the location of all of the EV chargers in Europe.

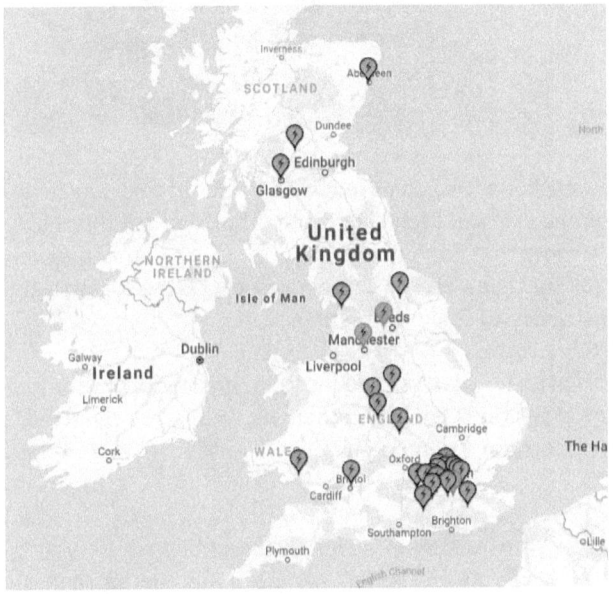

If you look at the map above I think that this is what some people think that the UK charging infrastructure looks like. This screen grab from Zapmap shows just the Shell service stations that have EV charging points. What charging points are there at a particular location? Let's have a look.

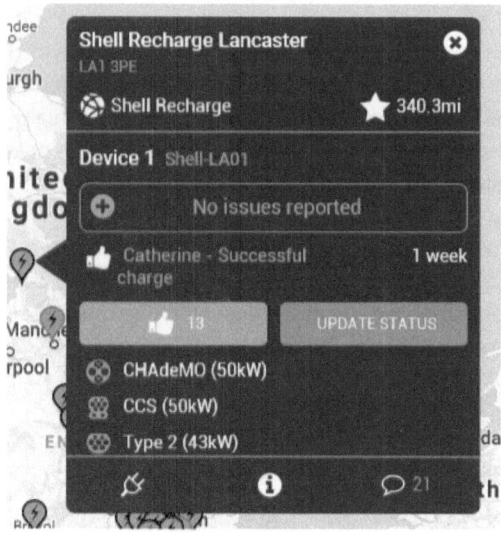

If you click on the coloured marker you get a small box of information. If you then click on that box you get this. I actually didn't know that Zapmap did this. As you are driving towards a charger someone else in the car can check the location before you get there to see if everything is OK.

On this page you can see the location, postcode, of the charging station. The condition of the chargers, 'No issues reported,' and in the photo on the next page you can see the actual charging point.

This one is positioned in a small parking bay where you pump your tyres up; it's on the left hand side, as can be seen by the windscreen and bonnet of my car.

My daughters Nissan Leaf has a type 2 charger and as can be seem, this station charges at up to 43 kw. This compares to the less than three that she gets at home.

If you are travelling up the M6 through Lancashire and need a top up, and have missed either Burton in Kendal if you are travelling south bound or Forton if you are travelling north bound then turn off at junction 34 and head towards Lancaster. If you use the post code you need to continue towards Lancaster for another quarter of a mile. Basically you go past the Holiday inn which is on your right hand side. Then there is a McDonalds and just around the corner is the Shell garage.

Unfortunately, as of yet, there is not a universally recognised EV charging point road side sign. This would make it easier to spot these charging points.

This is one of the common comments that I get from people. They say, "But there aren't any charging points near me." I can show them their area on Zapmap and they will be shocked to see how many there are near them. They didn't know that they were there because they hadn't seen them.

I want to show you just how many charging points there are currently, June 19, in the UK.

I won't ever let anyone say to me again, there is not a big enough infrastructure in the UK for electric vehicles.

One of the things that I was really impressed with at Fully Charged Live was that Shell, yes the petrol station people, have now added EV charging points to half of their forecourts. Because a lot of charging points are not obvious, people don't 'see' them. The more supermarkets that have them and the more petrol station, the more people will see them and realise that it will be easy to top up their electric car.

Let me tell you about the charging points near me. I live near Chorley in Lancashire. I know it is not the centre of the universe like one of the big cities but I am impressed with its charging infrastructure. In the town centre there is a car park called the

Flat Iron. This has four charging points. Then, the other side of town there is ASDA, they have four charging points too.

Eight charging points might not sound a lot for a population of 40,000, plus the surrounding area, so let's see how often you would need to use one of these.

It is estimated that most UK motorists travel about 15 miles a day. That works out to be 75 miles a week. If you are driving a relatively new EV then you will be able to drive for two weeks between top ups.

Really new news and a couple of last thoughts - Norway has just announced that from 2025 no new Petrol or diesel cars, including hybrids, are going to be available for sale. This is going to force the garages, which currently sell Petrol and Diesel, to change their forecourts to include fast chargers and hydrogen filling stations. Germany and Holland are doing the same in 2030. I'm wondering when the UK will be announcing the same changes. Most ICE cars last for ten years or more, so they will still be available for a while, but I do foresee a time when getting hold of fuel will become as scarce as charging points were in the early days of electric cars.

The UK currently is stopping the sale of new Petrol, Diesel, and apparently Diesel trains, in 2040. There is thought a campaign to have this reduced to 2032.

On my drive into work this morning I was thinking about the video that I had watched the day before on hydrogen fuel cell cars. As the Hydrogen passes through the cell, it attaches to oxygen in the air around us, and produces water but, giving off energy that it turns into electrical energy. If we attach that to an electric car that has a battery with a capacity sufficient range of 20 miles, the car then has the ability to power itself. It can recharge the battery by both regeneration and from the fuel cell. It is a really good idea as it is quicker to fill up than a pure electric car and, with a sufficiently large filling station infrastructure; it would relieve both range anxiety and the perceived problem of having to wait to recharge the battery. Again, you only need to charge the car up if you have driven sufficient miles to empty the battery that day because every morning you can start off with a fully charged car.

Then I had another thought. I know that you can have a car converted to be powered by LPG, Liquid Petroleum Gas, but can you power one on hydrogen? It is in effect, just another combustible gas. Some companies have played around with hydrogen enriched air-gasoline engines. You set the fuel mixture to extra lean and introduce hydrogen. This produces a slower and cooler burning engine with improved efficiency, but if we want to get rid of hydrocarbon fuels totally, then this technology, as far as I am concerned, is too late. I must, though, give credit where credit is due and give them a pat on their back for ingenuity.

For a hydrogen ICE car, I would assume that some of the internal workings of the engine would have to be changed. Firstly you are introducing a gas and not an atomised fluid. I don't know what the burn rate difference is between petrol and hydrogen; this could affect the timing of the spark. I don't know how hot hydrogen burns in comparison, so I don't know if there are any cooling changes to be made. What I do see though are possibilities.

You will notice that I haven't mentioned that hydrogen could be a replacement for diesel engines. Diesel fuel oil has a flash point of between 52 & 96 ºC. Hydrogen, on the other hand, has a flash point of 500 ºC so is incompatible for use in a compression-only ignition system.

So where do we get hydrogen from? There are currently two sources; firstly it can come from the oil and gas industry where it is a by-product. The other source is by creating it. Maybe creating isn't the right word. Water consists of two parts of hydrogen and one part of oxygen. We can use electricity to breakdown the electric bond that keeps them together, let the oxygen escape and just keep the Hydrogen. There is a company on the Shetland islands that are manufacturing hydrogen plants, the size of a small container, that is powered by either solar or wind, that carries out this separation and then pressurises the gas that it produces ready for topping up the tanks of hydrogen cars.

Here are two major differences between electric cars and hydrogen cars. Batteries are expensive but charging is cheap, so there is high initial outlay but low operating costs. Hydrogen cars will be cheaper to buy but, will cost around the same price to run as a conventional car.

Hydrogen power cell trains and trucks are a great use of the technology, and there is generally space to store the compressed hydrogen.

Chapter 4, Electric Bikes, Scooters and Motorbikes

If you can ride it, you can electrify it, and with the high energy batteries available today, the speed and range are comparable with any other form of two-wheeled transport.

Let's start by looking at electric bikes, or EAPC's as you can find on the .gov website. EAPC's stands for ' electrically assisted pedal cycles'. I found these details on the 29th March 2017, so recheck for yourself before buying or using an EAPC as the rules could have changed. You have to be over 14, so instead of waiting for your first moped at 16 you now get power assisted travel two years earlier. To ride an EAPC you don't need any licence, nor do you require being registered, taxed or insured. Legally you don't need to wear a helmet. Whether you wear a helmet and have your bike insured is up to you.

A rather interesting side note that I had not thought of is that EAPC's can have two wheels or three. They can even be a tandem.

Any EAPC that you are thinking about buying and riding on the road in the UK must be Type approved. To find out what that means, check out the .gov website. I would assume that any high street dealer such as an auto factors like Halfords or cycle shop like Charnock Richard cycles near Chorley, would only sell approved EAPC's. The electric motor can only assist you up to 15.5 mph, and the motor is limited to 250 watts. There are obviously other requirements, but they fall outside the scope of this book.

A pedal assist bike does exactly what it says on the tin. You turn the pedal, and it assists you. So where can you ride these bikes? As long as you have followed all of the rules, anywhere. Early electric bikes were just battery powered but with the changes to all of the technology recently, they too, now come with regenerative capabilities. Battery range will be limited by how much assist you want to apply and how much you use it. Price wise, I've seen these start at £1,000 and go up to as high as you would like to pay.

So, what about if you are a keen mountain biker who loves going up and down the trails but want to spend more time coming down than pedalling up? You can add more power, and I mean a lot more power, and you can have it so that you don't need to pedal. But, and it's a big one, you can't ride it on the road. It's like off-roading on a motorbike. As long as you don't go on public roads, you don't need to have passed your motorbike test, have an MOT or any of the other legal requirements of a road motorbike.

There is a video a YouTube showing a guy who put a motor in each wheel hub. He gets up to speeds of over 60 mph.

If you already own a bike that you love riding, you can now buy kits that consist of a front wheel with a motor in the hub, a battery, throttle for your handlebars and everything else to turn your bike into an EAPC. A quick look on Google and I came up with a couple of kits for less than £200. I don't know if they have the regenerative capability of the factory produced bikes, and I did notice that some of these kits didn't include the battery which could cost you another £200. A brand new electric bike, on the other hand, will cost you between £2,000 and £3,000 depending on what style you are looking for. If that is a bit pricey, you can see some for as low as £999.

Assuming that you are like me and no longer want to pedal something but you haven't got your motorbike licence and don't want to be bothered with your CBT (Compulsory Bike Test). Your next option is the speed limited scooters. The first time I saw an electric scooter I was in Egypt. You could hire them to run up and down the prom. A few years later I saw my first one in the UK. This one though was quite a bit slower than the one I had seen in Egypt. They are limited to 20mph, so, therefore, they are not classed as a motorbike of any description. I have seen these ridden on the road in the UK, but I can't find any details on them on the internet. If you come across one and fancy buying it, make sure that you check up on the legalities.

An interesting side note. Learner legal electric motorbike requirements state that it should not have continuous power of over a certain limit. Current UK electric motorbikes that can be ridden with L plates are a lot faster than petrol 125cc motorbikes.

It almost seems to me as if all of this technology has just crept up on us out of nowhere.

As part of the research for this book, I went to Glasgow to a two-day conference called 'All Energy.' I hoped to learn a lot more about renewable energy for the home, but most of the companies attending had industrial size equipment, rather than the small home use ones so after a few hours I left the conference and didn't return for the second day.

As I was in Scotland, with my daughter, we decided to do some site seeing. We headed over to Falkirk to see the famous wheel that transports the canal boats from one level to another, a height of about 100 feet. The incredible thing about this huge device is that it only uses the equivalent energy as boiling a kettle to transfer these boats up and down.

While we were there, we met a couple who were on electric bikes. I asked them how they found them. Their reply was easy, brilliant. Since buying them twelve years ago, they reckon they have covered between 3,000 and 4,000 miles. I'd probably say that they were about ten years older than me; this would put them in their 60's. They looked fit and healthy which was not surprising because the landscape around the Wheel was very hilly, hence the need for the Wheel.

I asked them what it was like to ride these bikes. "They take the strain out of the hills," said Martin Ellis (If I remembered his name right). His lovely lady, whose name I didn't get smiled at me as she said "It makes going up hills so much easier, we even pass people with all of the cycling gear on. They do give us such funny looks."

Martin told me something that I didn't know, and is a great tip for you. I thought that the pedal assist part would only power you up to 15.5 mph, but he mentioned that if you went faster than that, i.e. downhill, then a brake applied to slow you down, restricting the top speed of the bike.

I don't know if this is still true for bikes today or was just the regulation for the bikes when he bought them. Anyway, his great tip is, with his bike having six gears; the brake is only activated when it is in 6th. If there is a steep hill, he backs it off into 5th and goes down the hill at whatever speed he wants.

Martin also mentioned that for any off-road use, he recommends that people buy bikes with wheels bigger than 24". The small ones are great for folding bikes and riding through cities, but if you try taking them off road, they can become quite dangerous.

The next level up the scale of motorbike, is the learner legal moped. So far, I have only seen these in scooter form. They are road legal for a learner to ride at 16.

There is a large selection of both motorbikes and scooters for those with a full motorbike licence. One of the companies that amazed me when I was looking for these was Harley Davison. I thought that they might strongly stick to their traditional long stroke engines, but no, they can obviously see which way the trend is going and want to ensure that they are still in the market when people are looking for eMotorbikes.

Speed wise, most of these are capable of 100 mph, but obviously, at these speeds and with the high drag of an un-aerodynamic rider, this will soon burn through the available electrons. Keeping your speed down to more sensible limits will allow you to ride for around 100 miles for most of these bikes.

Charging wise, most of these mention charging at home times. I would like to see them with fast chargers at car charging points. Adding 50 miles on in twenty minutes would be ideal. Again, as with the cars, this technology is in its infancy.

Having said that there is already a team who have ridden a motorbike all of the way around the world.

Chapter 5, Houses

As with all things in life, there are many ways to help you save money with your house. You can use cheaper electricity, you can use less of it, or you can do both.

There are a couple of companies out there that will provide you with a house full of LED lights for your entire house. The only one that I know much detail about is Utility Warehouse. Find someone in your local area, or where you work, that is a Utility Warehouse Distributor and ask them about 'Daffodil Bulbs'. UW Distributors are people who have a second business and have a vested interest in you. They will help you lower all of your utility bills and, as a member with four services (Gas, Electricity, Telephone and Broadband), you will be entitled to having all of your household bulbs replaced with LED lights. I have had my services through UW for over ten years. I could give you all of the spiel, but I will let your local distributor do that. I have been a distributor for a number of years. I am only a card-carrying member these days. I'm not looking for customers which is why I suggest that you find someone near you. Also, if you have any questions and issues, they will be on hand to help you.

So what is the difference between a traditional filament bulb, and an LED one? LED's or Light Emitting Diodes are cool to the touch, unlike the hot older filaments that they replace. They are generally made from plastic, so that if you drop them, they don't shatter and cause cuts. There are several more advantages, but the biggest one, as far as this book is concerned, is their energy usage.

A traditional bulb would consume 60 Watts. If you had a big room, you might have a 100 or a 120 Watt bulb. With LED's, their energy consumption is drastically reduced. A 5 Watt bulb would replace a 60 Watt filament bulb. Depending on how many bulbs you have in your house, you could have them all on and still only use the same amount of electricity as two or three of the old bulbs. As one last benefit, LED's last for thousands of hours.

So, what else can you do to your house to reduce your costs? If it is an older house, one that just never seems to be warm, or where the walls feel cold, then cavity wall insulation might be the answer. When houses were built in the '40s, '50s and '60s, they were constructed with parallel walls a couple of inches apart. It was thought that the single layer of brick would be warm enough and that the air gap would help prevent condensation and keep the house in a better condition. Well, and I can tell you this from personal experience, that these houses could be really cold. I grew up in a pre-war Semi that had single glazing and cavity walls. When I was in my mid-teenage years, we had the walls filled, and it made an instant difference. We also covered the insides of the windows with a plastic film, and the bedrooms were positively warmer.

The government is desperate to achieve its lower greenhouse gas emissions figures by whenever it has agreed to and is offering grants for cavity wall insulation. Check with your local companies, but this is what I read locally. The government will give you so much towards the cost, and you have to pay the rest. Let say for example the grant is £175 and the company doing the work quotes you £200; then you just pay the remainder. You can lower this cost if you and your neighbour want it doing at the same time. For example, my old neighbour and I were going to get our halves of the semi that we lived in cavity wall insulated at the same time, and it was not going to cost us anything over the grant.

Again, it depends what the deal is where you live and the size of your property, but even if it costs you a little bit, you should soon recoup the costs.

Double glazing. Do you remember all of those annoying phone calls that we used to get? I suspect that these days the vast majority of houses have double glazing. There is a huge variety of standards between companies. I've heard of people having window units falling out and of other peoples who weren't any better than the single windows that they replaced. Hopefully, most of that is behind us but fitting decent quality windows will really make a difference to not only your comfort but will help to reduce your bills.

Double glazed windows with glass that reflects solar radiation to prevent rooms from becoming too hot. Also prevents heat loss from the room. The space between the panes can be filled with argon to prevent heat loss further. A lot of double glazing units in the '80s were aluminium. The glass was doing a great job to prevent heat loss, but the aluminium edges were transferring a lot of heat so look for units with warm edge space bars. If you want to reduce the cost of your energy bill, then looking into high-efficiency double glazing units is a must.

The only downside is, as far as I am aware, there aren't any grants for double glazing.

Loft insulation. For all of the same reasons that you should get your cavity walls done, you should add several extra inches to your loft insulation. If you don't think loft insulation is important, go and get your head shaved and see how much you feel the cold, and how cold you get as all of your heat dissipates through your head. OK, don't get your head shaved, but hopefully, you see my point.

There is a grant for this too. I'm not 100% sure what it is but if your loft insulation is below a certain thickness, there is money available to you.

So we have reduced the amount of energy that we use. So now how can we reduce the price of the electricity that we use?

Solar panels, heat pumps and underground heating. Let's look at these in reverse order. The ground under our house is warmer than the air above it. You don't need to go down too far to find heat radiating upwards. This type of energy source is good for a self-built house in its own grounds. You dig a large deep trench and lay coiled pipes in it. These are then covered over, filled with water and pumped through the house. Well, that is the basics of it. If you are intrigued, then you will need to do some research on it. It might be suitable for smaller homes with just a back garden. There might even be grants for it, but I haven't got any other details for it at this stage. I only mention it in this book to try and show the reader every possibility that there is out there.

I will quickly mention log burners. A little random in a book on electricity I realise but stick with me. A log burner is not like an open fire. By controlling, or limiting the flow of air and therefore the amount of available oxygen, you can slow down the burning process. The log burner will remain hot all night, and will heat the house for a lot less in logs than a traditional fire. Also, and these I only learned about recently, you can produce electricity by burning things. The first ones I saw were little camping fire cooking stoves that you put some wood in. There is some electronic gizmo in the device that turned some of the heat into electricity so that you could charge a phone or tablet while you were camping, and cooking your tea.

Heat pumps are strange things, and it took me a while to get my head around the technology. Essentially they are a fridge in reverse. The air in your fridge, at 8 °C, has some spare heat in it.

The refrigerant moves through your fridge and draws out this heat until your fridge is down to 5 °C. This heat turns the liquid into a gas. As the gas is then pumped through the radiator, at the back of your fridge, it turns back into a liquid by giving up the heat. Feel the back of your fridge, and you will see what I mean. An Air Source heat pump is four times more efficient than a pure electric or gas system, and using ground source; you can get up to seven times as much efficiency. This means that for 1 kW of energy used, you will get seven kW of heat.

When I understood how the heat pump worked I realised that we had been using similar systems on aircraft since the 1940s, but in a slightly different way. I do apologise and say that this is about to turn into a physics lesson, but if you want to understand the principle then read on, if not skip to the next paragraph.

To produce the air for the cabin pressurisation in a passenger aircraft, air is taken from the jet engines after it has been through several stages of compression. The first thing that I should mention is that aircraft fuselages aren't pressurised like pumping air into a cylinder. It is more like putting in a lot of air and restricting the air going out. There are regulations controlling how much fresh air each passenger must get.

The air from the engines is quite hot, say around 70 degrees C. The air flow is split into two, one stays hot, the other one is compressed further by a compressor fan then passed through a heat exchanger where outside air cools the compressed air that is now a couple of hundred degrees. This cooled air then passes through a turbine taking out more energy to result in very cold air. The hot and cold air is now mixed to produce the cabin air.

Let's look at this in more simplified terms. If you have an empty 2-litre pop bottle that was full of air at room temperature, say 20 °C, then you compressed it to half its size. The 1 litre of air would now be hotter than it was before. I don't know that if we half the volume it would double the temperature inside. You would need to ask a fluid dynamist that question. But I think that it is fair to say that it would increase it considerably. Then passing that warm air through a heat exchanger (think of a car radiator), the passing air would warm up, and the internal air would cool down because nature always likes to find its balance and move from hot to cold. Incidentally the higher you compress the air, the higher the temperature differential, the more heat will be removed. Therefore the more efficient the device will work.

This is similar to how the heat pump works, and why they can take in air at something like -10 ºC and actually pump warmth into your house. I just recommend that you don't stand anywhere near the outlet, certainly on days when it is already below zero.

Let's talk about wind turbines for a moment. If you have a house on a large plot of land, say a couple of acres, you certainly can put up a small wind turbine. As the wind blows, it will generate some lovely electricity for you. Once the blades are up to optimum speed the bladed will turn at a constant rate and the stronger the wind, the more kW's you will generate. If you were to combine this with solar panels and a large house battery, you could charge up your fleet of cars and farmyard equipment. There are currently electric sit on mowers and even prototype all-electric tractors.

If you live in a house, on a housing estate, then it is quite a different matter. I personally would love to see windmills on all houses but, unfortunately, at this moment in time, they are not practical. Not from what I have read anyway. The problem with windmills is scale. On the farm, they would have a ten meter three bladed unit. I came across one of these in Germany in the '80s. Since changing over to wind power, he had never had to take anything from the grid.

A small say 12v 500W wind turbine would cost you around £200 to £250. This is great for caravaners who want to top up their leisure battery, although these days a lot of caravans use hook-ups. It might also do to keep a couple of batteries topped up in your summer house (shed) for the nights when the barbeque turns into a late one. To be honest, these days, a small solar panel would do just as good and cost a lot less.

One of the issues is that as you make a windmill smaller, it produces a lot less power. Blades cover a small percentage of a propeller disk, around 18% is a good starting number. So let's look at two wind turbines of 3ft (1m), and 6ft (2m) and, using the formula of Pi r^2, let's see what 18% of their area is.

3ft (1m) r= 18" (0.5m), 3.142 x 18^2 x 0.18 = **183in²**
(3.142 x 0.5^2 x 0.18 = **0.141m²**)

6ft (2m) r= 36" (1.0m), 3.142 x 36^2 x 0.18 = **733in²**
(3.142 x 1.0^2 x 0.18 = **0.566m²**).

You can see by the figures above (assuming that I got them right), that the larger turbine has four times the area of the smaller one.

I certainly do hope that in the near future small, efficient home turbines can be developed. I know that the apex of your roof has a lower pressure than under your eves. Is there enough pressure differential there to make a device that can capture and turn into useful energy this difference? I hope so.

House batteries. I briefly mentioned these earlier. What is a house battery? It is a small box that fits on the outside of your house, or inside your garage. As you produce an excess of electricity, solar or wind, during the day, this battery stores it. Ten years ago, the only way to store this energy was by using Lead Acid batteries, and you would have needed a lot. The Tesla House battery can store 10 kWh, more than enough to last you a few days. A house battery also means that you can be either independent of the grid, or you can actually supply to the grid, but you will never be without power should you experience a local power cut, (This does have some other requirements).

Redflow is another company making house batteries. They currently have two batteries, one that stores electricity in a solution that it pumps around. This is called a flow battery. Instead of storing the energy in lots of batteries, which is actually quite safe due to the low energy usage per cell. The fluid in this battery is the only part that gets energised. Any heat that is built up during this process is distributed into the fluid so that the whole battery remains nice and cool.

The other is called a Zinc Bromide Hybrid Flow Battery. To store electrical energy it zinc plates the plastic membrane. To release the energy it un-plates the sheets. There are 32 sheets in this house / small business battery that has a capacity of 10 kWh. Unlike Li-Ion batteries, this can be fully charged and fully discharged down to zero. But, the energy release is slow so is ideal for home or office application but too slow for a car. There is not much risk of fire as it is not an energy dense solution, like the cell batteries. Actually, bromide has fire retardant properties.

For larger installations, like one that they currently have in Australia, 32 house batteries are connected together in a test rig supplying an office complex of 80 people. The battery storage is actually sufficient to supply the office with full power for up to three days. That assumes that the main grid goes down and that the sun goes out!

This is a completely new system, especially their household Z Battery. If any electricians want to be part of it, then they can contact Redflow to see about becoming an authorised installer. Have a look on the installer's page on www.redflow.com.

Chapter 6, Other Home technologies

I've already covered transport and producing electricity, but what else can we expect to see in the home?

I remember electric lawn mowers in the '80s. They were very limited in power and, if you weren't careful, you could run over the cable with severe consequences. To alleviate this, they started producing battery mowers. Initially, the mowers were limited to 12-volt batteries, so they were very underpowered. Batteries in hand tools have been steadily increasing since then going to 24 volts and up to 48 volts.

Why does voltage matter? Effectively, voltage is pressure. It is described as 'Potential difference'. One side of the battery might have 5 volts, and the other one has 17 volts, so the difference is 12. That is how much pressure there is behind the electron to get it from one side of the battery to the other.

Therefore a 48 Volt battery electric drill has a lot more power than a 12 volt one.

This sort of technology has also made its way to lawn mowers.

New, professional lawn mowers now have motors of 2 hp. That, and its associated torque, is equal to a petrol mower. This is coupled with two 3 kWh 48 Volt batteries and gives you over an hour and a half mowing time. This mower, and all of Bosh's professional range is designed for people at work and has a full range of accessories. Two spare batteries, two chargers and a battery block and you can mow all day without needing a socket.

In theory, you could have this pack charging from solar panels on the back of your truck.

What else is available for you that you could charge up around the house? Anything that you can think off. There are skateboards, personal flying craft. Single wheel, I don't know what they are called, but it's a single wheel that is powered and either side there is a foot plate. You stand on this, level it and away you go.

Really the possibilities are endless and are only limited by our imagination.

Chapter 7, Our Short Term Future

There are already, today, March 2017, electric aircraft! Some of the first experimental ones were microlights, that had the engine removed, batteries, motor etc. installed and they flew for about ten to fifteen minutes. They really looked like fun, but you didn't want to leave the area around the airfield, what pilots call, staying in the circuit. Although these aircraft were fun, the fact they got off the ground was amazing. But let me tell you what is already flying. In America, and this has been flying for a few years already, (I actually use it in one of my novels), is the Electra. I really like the Electra because it has a glider style cockpit. The wings are attached behind the cockpit, so you get an uninterrupted view, unlike most light aircraft where you are sat on, or under, the wings. It has a tricycle undercarriage arrangement so that it can taxi itself to the runway and is, therefore, a single person operation. Most glider operations require several people to organise an aircraft launch. In its motor glider format, it has a two-hour power on, duration. In its light aircraft configuration, this is doubled to four hours. It is light and has glider-type wings and is designed to soar like a glider so these flight times can be dramatically increased. It is also FAR103 certified, which is the American, single seat homebuilt experimental class of aircraft so that you require a medical not a pilot's licence to fly it though, training is always recommended.

If you want to learn to fly in the UK or Europe, there is a two seat side by side aircraft that I think is about to go on sale. I'm not sure if they have finished all of the certifications yet. This aircraft is designed to teach students in the circuit as it only has an hour power capability. Within this hour several circuits can be flown using cheap flying on electricity.

There are five costs involved in flying an aircraft: the instructor, landing fees, fuel, aircraft hours and engine hours. The two largest parts of this are the amount of fuel that is used and the engine hour cost which is money put into a pot to pay for the expensive servicing of the engine every few hundred hours. The aircraft in most flying schools use Lycoming and Continental engines. Modern aircraft and Microlights use Rotax engines that are far more modern in design and more fuel efficient than the old engines. Lycoming and Continental engines date back to the 1950s and 1960s. This is mainly due to the requirements for new engines and the associated costs being so high. So these engines used leaded fuel that is 100LL. Today at the pumps for cars, so-called mogas (Motor Gasoline) we have ordinary unleaded and super. Before the '90s these would have been two star and four star. The increasing star rating denoted the higher energy available. The 100LL that these aircraft use is equivalent to the old five star. So you can see that not only are they using out-dated engines but they are also using outdated fuel that contains lead.

So, back to our two-seat aircraft. The propeller is designed to increase drag when it is not in the power mode. Most aircraft deploy flaps from the trailing edges of the wing to help increase drag so that they can land at a steep angle. This makes landing easier. On this aircraft, the propeller provides this service and therefore turns faster when the power is reduced. This increase in the rotation and with no power coming to the motor, it is now turned into a generator. So it takes something like 5 kWh to gain 1,000 ft in the circuit, but it reclaims 2 or 3 kWh as it descends to land.

This for me is one of the most significant parts of the technology. It doesn't matter if you are riding a bicycle, riding a motorbike, driving a car or flying an aeroplane, if you can reuse braking energy by storing in a battery and not waste it in warming up brakes, then you dramatically reduced your energy usage, and electric motors are already super-efficient anyway.

A few years ago I bought myself a Paraglider and started taking lessons at the weekends to learn to fly it. I soon got bored and knackered with all of the walking up hills. It's a bit like skiing. You spend five times as much time walking uphill as you do coming down. That is why, at ski resorts, they have ski tows. In the paragliding world, they have paramotors. This consists of a small petrol engine, about twice the size of a lawnmower engine, and a propeller that is carried on your back. I bought myself one of these, and although it was fun, it was messy and didn't always start. It was like the old aero model engines. They were always difficult to start. The electric radio controlled aeroplanes are so much easier to use. There are now electric paramotors. So we have today an aeroplane that packs into a backpack, a motor that just sits in the boot of your car and batteries that you charge at home. At the weekend, given the right weather conditions you pack everything into your car, go to the flying site, set up, switch on, no fuss, no hassle and, if you have solar panels, you fly on sunshine.

People have also been experimenting with electric packs for Hang Gliders.

The one type of aircraft that really impresses me is the electric self-launching Gliders. Usually, gliders are launched by a winch launch or an aero tow. There have been motor gliders around for years, but these don't have very impressive glide ratios. Most of the early ones were basically light aircraft with extended, high aspect ratio wings. In more recent years there have been some more impressive self-launching gliders, and these have involved having moveable nose cones and flipping out propeller blades. They have also involved using large engines too. It is the dream of all glider pilots to have an aircraft that they can get from the hanger themselves and launch through a one-person operation or at least with minimal crew. I have recently come across one such glider. For the last twenty plus year's glider manufacturers have tried adding self-sustaining power units to the back of gliders with pop up masts that support a large propeller. This type of configuration works but the engines tend to have a high vibration, and if there is a problem starting the engine, the mast provides a lot of drag. The reason for the large propeller is due to the slow turning of the engine.

The electric powered glider that I saw has had the front of the fuselage removed and replaced with a spinning nose, which is less than a foot across. Attached to this are spring loaded propellers blades that sit tight into the fuselage in flight, but underpowered flight are pulled out by centrifugal force. The battery life is ideal for a glider pilot. There is enough power for a single launch to 5,000 ft, two launches to over 1,000 ft, or a 1,000 ft launch and half an hour sustaining flight power if they are struggling to find lift. This can mean the difference between landing back at the home airfield or having a retrieval crew extract them from a farmer's field or the bottom of a hill.

While you might not come into direct contact with any of these leisure aircraft, you will, no doubt, come across the following. Studies are currently being conducted with commercial passenger aircraft to see how they can help the environment by saving fuel.

At quiet regional airports it is not unusual for an aircraft to taxi to the end of the runway and take off, but where you have busy runways, any slight delay can cause a traffic jam on the taxiway and for there to be several aircraft with their engines running sat there burning fuel. One option is to have motors on the main undercarriage wheel units (bogies) that power the aircraft towards the end of the taxiway. The motors and batteries can have a significant weight, but it is not unusual to burn off a ton of fuel while taxiing. At busy airports, this can be doubled or tripled.

There are also studies being conducted as to how solar panels and motors could assist the engines. Aircraft require the most power at take-off. If some of this can be supplemented by electric motors, then that will assist in fuel savings. When aircraft are flying at cruise altitude the amount of electricity that the solar panels can produce is increased by the fact that the sunlight doesn't have to pass through as much air. A moderate thrust from a couple of electric motors could again help reduce fossil fuel usage.

During the studies, it has to be determined if a particular aircraft type would actually spend enough time flying in daylight to help offset the additional weight. I look forward to seeing what creeps into the commercial flying sector.

One last thing while we are still on aircraft. Have you ever watched the Redbull racers? The ones that fly around the inflated cones over lakes and rivers? Well, there are a couple of new aircraft that will be electric motor battery powered. They are ideal for these types of flight as a typical routine flight lasts only fifteen to twenty minutes.

A prototype version of this aircraft has taken the absolute speed record in both the under 1,000kg and over 1,000kg class. I assume that they just fitted enough ballast to make its weight high enough to fly in the higher category. The following day it towed a glider to a height of 1,600 feet in just over a minute. That is an incredible feat.

I was reading last week in the news that oil tankers have started sprouting large pole type masts. These usually consist of two at the front and two at the back. It was found in the early 1900s that as the wind passes a rotating cylinder, it causes a low-pressure area and therefore provides thrust. Large tankers have been found to save up to between ten and twenty-five per cent of their fuel. Considering the amount of fuel that these vessels use, that would be significant.

All of this fuel saving in aircraft and ships will help to allow the fossil fuel we have left, to last longer.

But what happens if you want a yacht to sail around the world. I don't mean one with sails, I mean one with an engine, or should I say motors. There are several available at this time. In 2012 one sailed all around the world at the cost of no fuel. With the batteries fully charged one that I read about has a battery-only range of 72 hours at 9 knots. While 9 knots might not seem to be fast, it is a lot faster than you can guarantee a boat with sails will go. I could just see me cruising around the Caribbean in one of those…

Lightyear One
In 2010 Nissan launched the Leaf. I think the Renault Zoe was launched around the same time. Back then they were looked upon as somewhat of an oddity. While they were still proper cars, their range was severely limited. It is from these cars that the term 'range anxiety' was created. Just like you can't use a drag car for long distance touring, you couldn't use either of these cars for long distances without stopping a lot. However, if you are visiting a lot of places in a relatively small area and have arranged to recharge at each stop, they it wouldn't be noticeable how short its range was. It's the same as if you live in a city and never go out of that city, you work there and all of the shops and places that you visit are there then again you will never have a problem. The only problem with these early cars was that you had to keep on top of the amount of energy that they had stored.

The Lightyear One has had its first rolling outing, but it promises a lot. This is the first preproduction model that the team have built. It is not a concept car, like some teams build and show, that will never become a reality, the One will be on the roads in the next year or two.

So how does this 2019 car compare to the earlier EVs that I mentioned above? The One is up to twice as efficient as other EVs and in addition is solar charged. It is calculated that its high speed winter range with car heating on etc. will be about 400 km 250 miles, which is assuming no sun so very little to no solar charging. Its summer, solar charged daily range will be up to 900 km (560 miles).

While you might never want to drive that many miles in a day, what it could mean is that from May until September you never need to plug it in. the company is based in the Netherlands so all of their figures are quoted from there but they estimate that they can get an additional 6,000 km (3,700 miles) per year from the solar where as in California or Australia they will be looking at up to 20,000 km (12,500 miles) per year.

If you do need to plug it in, a European outlet (240 V, 3 amps) will increase the range by 400 km (250 miles) over night and it has the capacity for fast changing.

Solar charging is up to 1.25 kWh and can add 10 km (6 miles) per hour. That is not a lot different to plugging your car into a standard socket at home.

So from having a car that does only 100 miles per charge ten years ago to a car with these sorts of stats, is amazing.

If we take its maximum range, 900 km (560 miles) and divide it by 24 hours this car could travel at 23 mph for a whole day.

The One is currently on pre order in the Netherlands for £133,000. It is defiantly the high end car. They are planning on making more affordable cars for the general public but as is the current model you can make a better name for yourself, and make more money by producing less high end cars to begin with, just as Tesla did with the Roadster and the Model S.

So who is this team and how can they claim to be so good? Have you ever seen these solar races across Australia? Well this team won it. I can't remember if they said that won it more than once. So they know about making things more efficient.

In this car they have the motors in the wheels, this reduces weight inside the car. Because they have a more efficient system, they have less batteries and therefore less weight. Less weight and regen requires smaller brakes. Less weight in the entire car reduces the amount of inertia that has to be dissipated in a crash, and so it goes on.

I am really excited to see this car on the road.

Chapter 8, My Ideal House

I would want my house to have the best insulation that I could possibly get. This would reduce my energy requirements for both the summer as well as in the winter. I would have as much natural lighting as I could possibly get, and use the lowest energy bulbs that I could find that would give me the lighting quality that I have grown to expect.

I would have solar panels enough to fill the side of my roof facing the sun. I get paid for every kWh that I produce, so I want to produce loads. I will have a commercially produced car as my run around; it will be the most energy efficient. My second car will be a seven series BMW, an old one. I prefer the older style ones, the ones with the car phone in. Weren't they so high tech back then? Why the seven series? I used to have one. I walked to work, so I didn't waste any money commuting, but it was just so luxurious to drive. Big cars like that have amazing suspension. The designers at BMW certainly know how to make a comfortable car. I need to find one that has a worn out engine then get one of the new car conversion companies to make all of the necessary modifications. To drive it, there will be no differences from the one that I had, except that it will be quiet with only the hum of the motor and it will run on sunshine. I will also have an electric motorbike for all of the similar reasons. I love riding around the bends of a country road on a nice warm summer's day. I won't be getting myself an electric bike. It's just not my thing. An electric boat on a lake or the sea would be good, but I really fancy an Electra down at my local airfield. I want to be able to cook on gas for both my home cooking and barbeque on hydrogen that I have produced myself and to cut my grass with my powerful electric mower.

It is my dream to be able to live my normal life for a whole year just on energy that I have produced. That is without using anything produced by national companies, only energy produced within the grounds of my house, but to still be able to go and do all of the things that I love to do.

So here is a question. If we are using all of this solar power, will we ever use up all of the sun's energy, like we are doing with fossil fuels? A lady in California actually posed this. The easy answer is no. Firstly, the sun gives us energy, but we don't take it. We are just utilising what the sun has covered the earth with for the last few billion years. The long and scientific answer is that is it hard to imagine the full size of the sun. Let's compare the size of the sun to the size of the earth. Actually, that is quite difficult. Let's then compare the size of the sun to the distance between the earth and the moon. Our moon is at an average of 225,000 miles. If the earth were in the centre of the sun, the moon would be only just over half way across. The sun has a radius of 400,000 miles.

I actually did some interesting calculations a while ago. I'd love someone to check them for me. I calculated that the piece of the sun that shines towards us has a diameter of 40 miles. Using Pi r^2, it gives us an area of 1,257 square miles of the sun's surface that provides all of the sunshine that the earth gets.

Solar Home System
I'd now like to show you something that I bought a while ago.

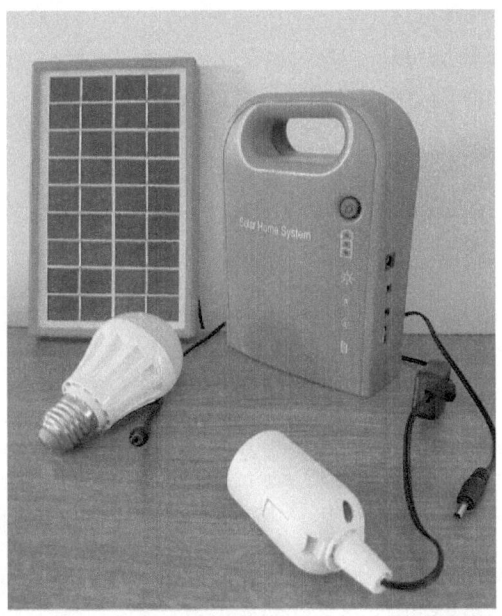

A girl in Canada developed a hand energised torch for her friend who lived in Fiji so that she could see to do her homework. Imagine what sort of difference this kit could do for millions of people around the world.

When I bought it I think it was £15. The last time I looked it was £50. I don't know what these cost to make, probably a half or even a third of what I paid.

No-one should ever be in in the dark or not be able to do their homework and get a decent education because there is no light.

Chapter 9, Our Long Term Future

I don't know if there are any gains to be had in electric motors. They have been around for a couple of centuries now, and I know that they are already super-efficient. An electric motor doesn't have to compress anything like a piston engine or a gas turbine. All it does is rotate on ball bearings. Because nothing else touches, there is nothing really to wear out. Old motors and alternators used to have brushes. These were solid carbon contacts that connected the spinning contact to the wires. Most motors these days are brushless. What that means I really don't know, as I'm a mechanical engineer.

The two biggest changes are going to be in the efficiency of the controllers and the energy density of the batteries. Changing from lead-acid batteries to lithium-ion ones took us from the humble milk float to the Nissan Leaf. Work on these batteries over the last couple of years took us from the 130 mile Zoe to the 250 mile one, for the same space and just a little more weight. Where can we go next? There are several groups out there working on different types of battery mediums. These are generations ahead of the current batteries, and because of the incredible car sales, there is funding available. As I said previously, over half of the new cars bought in Denmark last year (2016) were electric.

Several companies are now working on solid-state batteries. Tesla's new Roadster has a range of 500 miles, what batty technology that is using I don't know.

If home and car batteries have ten times the storage capacity and solar panels and other types of renewable energy system are developed then there is no reason why homes should not become self-sufficient energy wise.

But I really like cooking with gas. Oh, I know what you mean. There is nothing like the controllability of a gas cooker. There is a way for you to still have this.

Once the solar panels have topped up our house batteries, we can use excess electricity to split water into hydrogen and oxygen. If we collect and pressurise the hydrogen, we can use it within the home.

But isn't hydrogen dangerous? One gallon of petrol (gasoline) has the same energy as seven sticks of dynamite. Gaseous hydrogen, on the other hand, has the same energy as domestic gas, or barbeque gas. You could pipe it into your house and use it to cook on. The only disadvantage is that it has a clear flame. We can see natural gas because it burns off the impurities. So there are two ways to use this: add something into it that burns or use ten per cent natural gas. I don't know if anyone is developing something like this for either commercial or domestic use. I could see solar farms producing gas when the demand for electricity is low. They could produce hydrogen and add it to the domestic gas supply system. This would be useful on a windy day, instead of turning off wind turbines they could top up the gas system and reduce our use of natural, fossil, gas.

Just to complete this last chapter that I am sure will grow dramatically over the coming months, there are also in the pipeline, companies that are looking at turning waste farming materials, like stalks, into fuel. Another one is looking at cultivating waste ground to grow high yield grasses for biofuels to turn into electricity to power cars.

Here is one that I would like to add. Other people might be working on it, and that is great. But the question is, what do you do if you run out of electricity in your EV? Well you shouldn't is the easy answer and I would agree with that, but what if the last three charging points that you have been to were out of order? Currently, you can have a range extender engine, but there are two problems with these. Firstly, we are back to burning fossil fuels again, ok, only a little bit in an emergency but still, there are fossil fuels in there. Secondly is the cost. Don't quote me on this but I am sure that I have seen figures of £6,000 for this little, essentially, motorbike engine with a generator on.

So let's tackle the energy source first, hydrogen. You can create hydrogen with electricity from water, all nice and clean. Secondly, you can create electricity by either using a fuel cell or an engine of some description.

Let's say you have an aluminium cylinder that is blocked at both ends. Inside is a piston with a magnetic core. Each end of the cylinder has valves and a spark plug. As the piston moves from one end to another, it passes within a copper coil, thus turning the motion of the pistons into electricity. This is effectively a free piston engine. These are really cool from an engineering perspective.

How do you get the gaseous hydrogen into your car? Here is the genius thought that I had today. Do you remember soda stream machines? They have a large cylinder inside them that holds up to 3,000 psi. From what I can find that is 60 litres of hydrogen. How far will that get you? I don't know. I would imagine that it would only get you 20 miles or so, but that would be enough to get you to the nearest charging point. If garages sold them you could at least produce enough energy to get you back on the road without waiting for hours. Also, breakdown vehicles could carry a few, just like they do for when people run out of petrol and diesel. Whoa, I figured that a few people every day would run out of fuel. According to a Daily Mail article in 2015, they said that over 800,000 people run out of fuel each year, and this number is rising. And you thought that you might have range anxiety in an electric car!

So here is my challenge to car manufacturers, large companies, small companies, groups of engineers and Bob in his shed in the garden. Produce something that can run on gaseous hydrogen from a soda stream type of gas cylinder. i.e. one that is currently available on the mass market, that will produce electricity at a voltage, ampage etc., that an EV's can use, that will fit in the boot of a car, that can carry away it's heat and any residual fumes via a simple modification to the car. Or something that you get out of the boot and place next to the car. All of this for a per unit, the full production cost of less than £1,000 or, ideally, below £500. Get your thinking caps on.

Having written all of that by 2020, there should be sufficient charging points around the country that you should never run out. Just have a look at Zap map to see the number of charging points that are available.

So what else is new? Well with the new increases in energy density, we can look at a man carrying multi rotorcraft. But let's look at energy density first. I mentioned earlier in this chapter that a gallon of petrol is equal in energy to seven sticks of dynamite. That is a lot of energy. But, of that energy, a modern engine only produces power equal to around 18%. Electric motors, on the other hand, are about 60% efficient. What powers electric cars are obviously batteries. A triple-A battery that you might use in a camera has up to 1,200 mAh. AA batteries, the best that I found had 2,600 mAh. Both the AAA and the AA are 1.5 Volt non-rechargeable batteries. The new Tesla 2170 battery, so called because it is 70 mm long and has a diameter of 21 mm. This is a 3.6-volt battery of 6,000 mAh. It is an improvement from the original battery that has 3,000 mAh.

As you can see, for only being slightly larger, these batteries have twice the energy.

There are several companies currently looking at alternative methods of storing electrical energy. It is predicted that over the next five years electrical energy density could increase by five to ten times. Even if it doesn't, we are starting to see some amazing inventions with the current batteries. Let me just compare that to the lead acid battery that powers your car.

A typical lead-acid car battery will give you 40 Wh/kg (Watt hours per Kilogram). The 2170 is 250 Wh/kg. That is 6.25 times the power density. Let us go back to our 1970's milk float and replace the 1,000kg of batteries. With the 2170 battery, we would only need 160kg. When future batteries reach ten times their current density that would reduce to 16 kg. Well, these batteries would open up a whole new world of possibilities, for the milk floats that would increase their load carrying capacity by another ton.

Bringing us back to today, let's have a look at what technology is currently available. Amazon has started flight testing of its autonomous delivery drone. Today some companies can deliver to you the day after you order it. What about if they could guarantee getting items to you within four hours? This has just been reduced to a two hour delivery time scale. There will only be a limited selection of items as all have to weigh less than five kg, and I suppose they will be things like office essentials, maybe baby milk, but things that are not normally immediately available to us. To get your product to you, an automated drone flies from the local Amazon Depot and lands on your back lawn.

What about an air taxi service? You arrive at your destination airport, and you don't relish the hour it is going to take you to get a cab and sit there as the meter runs up as you are stuck in traffic. Wouldn't it be a lot easier to jump into a single seat multirotor aircraft and be whisked directly to your hotel roof? It is funny. I have spent years honing my flying skills so that I could fly aircraft. In the last few years, quadcopters have emerged that have incredible self-flying capabilities. There are some legal issues with where people fly these, but a friend of mine uses his to take landscape pictures up in the mountains, and he gets some amazing shots. The camera has a higher quality range than my DSLR, it is gyrostabilized, and he can position it so accurately using his phone that he can take pictures that we were unable to consider taking two or three years ago. These single-seat multi rotors utilise this sort of control and are therefore totally capable of providing this service. Do we trust autopilots systems? Think about it. We have been using autopilot computers in airliners for the last fifty years.

I'm sure that you at some point in your life will have watched an American film and seen a school bus. They all seem to be the same design, a 1950's style and yellow. Well, there is a company in Canada that has just produced an all-electric school bus.

So where else could solar energy be used? I would love to see cars that would directly charge from the solar rays. There is one that can charge just over two miles per day, which is good, but a bit limited. The problem is that at the moment, the only place to have the Solar panel is on the roof. I have seen solar panels that are made up of a few layers of film or that are sprayed on. This could be applied to all of the car and not just the roof, and should, in theory, give you eight to ten miles a day additional range, depending on latitude and cloud cover.

You don't always just want the energy at the wheels. I live in the north west of the UK, so most of the year it is cold or cool. Occasionally it is warm enough for me to use my Air Conditioning. I know that, in a fossil fuel car, the air conditioning unit can decrease your economy because there is both an electrical load applied to the alternator and a mechanical pump. What about if your solar roof ran all of your internal electrics via the battery? Your battery would always be fully charged, even if you had not driven your car for a while. Your headlights, AC and all of your other utility services would be covered by that battery and solar panels, until the battery got to the point where it required additional charging from the alternator. Unless you have an additional car battery for these services? Obviously for the electric cars all of their utilities already run off this battery. Again the solar roof would be of use here too. The Nissan Leaf actually has the option for a spoiler with a solar panel for the purpose of keeping that battery topped up.

I think that both trains and buses could benefit from solar panels on the roof, an electric motor and battery. Both of these forms of transport have to accelerate then slow down. If we look at this in a reverse sequence, the deceleration can charge the battery and then be used for the acceleration. All forms of transport have a propulsion system that is designed to cope with the highest demand that it is likely to see. An airliner, for example, has to be designed to take off from a hot and high runway so for most takeoffs it only uses 85% to 95% power. In the cruise, this is reduced to 60% to 65%.

I think that trains could have a hybrid system. A generator engine that tops up the batteries in the 'Engine' part of the train. This would stop/start depending upon the battery charge condition. As well as the 'Engine' having solar panels, each carriage could as well. These could either feed into the 'Engines' battery or each Coach could have its own motor and battery thereby relieving the strain from the 'Engine'. Buses and coaches could have similar adaptations. Anything that extracts energy from braking and applied it to acceleration saves both fuel and requires a smaller engine.

Why is a smaller engine important? Firstly, it should be more efficient, and it will weigh less. Why is an engine that weighs less so important? This is where I can really get into my particular discipline. I have been a weights engineer for twenty years in Aerospace, Automotive and the Marine industry. If we have an engine that weighs less, the structure that supports it can be made lighter. As the structural supports can run the full length of the vehicle, this can mean a thinner structure all over. The combined weight reduction of the chassis and the engine/gearbox combination can significantly reduce the energy reduction requirement for the braking system, so smaller brakes can be used, resulting in smaller wheels and lighter brakes.

This accumulative weight saving results in not only the reduction of weight but also in cost because smaller engines generally cost less. It also results in a more efficient system that costs less to run and requires a smaller fuel tank. I always find it amazing how a weight decrease, or a weight increase, can have such an effect of the rest of the platform.

I love American RV's. If you have ever enjoyed the freedom that a caravan brings you, you will enjoy driving a motorhome. If you have ever been on a motor homing holiday, you will know what it is like to stop in a layby that has an amazing view, turn around and going into the back of the Motorhome, put the kettle on and nip to the toilet. I find them so convenient. American style RV's are like a motorhome on steroids. It is like driving around in a luxury self-catering apartment. When you park up at the roadside, you can slide out of the seat and move into the back with ease. There are about four feet between the seats instead of a few inches. It has a proper porcelain toilet, a separate shower and a luxurious queen size bed.

Everything is full size, like the fridge, freezer, cooker, oven and in some, even the washing machine. When you arrive at a campsite, you connect up the electrics, change over the air conditioning to external instead of the engine run one. You connect the water and sewage pipes and pop open the valves and power out the slide outs. These increase your living area dramatically. You are now fully connected to the mains and can remain there for months if you want. Alternatively, you can travel on the road for a few days, or even up to a week with the on-board water supplies and water tanks. The only issue that I see with them is the size of the chassis and the engine. To feed the monster of an engine, the fuel tank is huge. One that I hired cost me $400 to fill, and that wasn't from empty! While retaining the luxury feel of the interior, a lot of weight could be saved. Added to this losing a ton or more off the chassis and having a smaller engine and half size fuel tank would result in a lot lighter vehicle.

With an electric drive system, a generator engine and a roof full of solar panels I would say that they could become reasonable economic vehicles, adding even more pleasure to the holiday.

I wanted to emphasise everything in that previous paragraph for two reasons. 1, with an electric vehicle you should not lose comfort, luxury or range. 2, neither should there be an increase in cost over the life of the vehicle. Most of the cost of servicing any piece of equipment is in the engine. Your hoover gets used more hours in a year, but your petrol mower costs you more to look after and is more liable to cost you in repairs. Annual servicing costs of your RV would be a lot lower as would your fuel costs. This would offset the additional costs of the initial purchase. I would hope that after ten years of ownership that you would be better off owning an electric/hybrid RV than a conventional one.

I want to go off topic here slightly and, I have to apologise for this, I am going to climb onto my soapbox.

Here is something that I would like you to consider. You can be part of the solution and not the problem. You can be the box and not the ferret to quote comedian Diane Spencer. In this country, we have a requirement for large power stations to supply us with the huge energy requirements that we demand from our electricity supplier. Most of them are old oil and coal-fired ones that in their day would have been efficient, but today, there as new standards would have been classed as inefficient, but now, 25 to 50 years on, they are wearing out and need replacing. The greenest way to achieve this was to design a new nuclear power station. These, at least, produce zero $co2$ emissions. There are, unfortunately, a hundred and one reasons why we don't want one around. Just think of Fukushima and Chernobyl. Then there is the fact that the fuel waste has to be stored for thousands of years and we are effectively asking future generations to look after our unwanted waste for us.

We are currently looking at building this nuclear power station. Now, I am not usually one to get on my high horse about something but with the current changes to the views on renewable energy, with the increase in wind farms, and more recently solar farms, I don't believe that we need this power station. So what can we do as individuals? We can use less energy, we can make our own using rule 1, if you can remember that from the beginning of the book and we can lobby whoever we can to prevent this colossal waste of money. I have no doubt that the cost today will double or even triple by the time the plant is finished. Then it will be the biggest and most expensive white elephant ever built. Next, you need to put yourself in a position where you can afford to own and drive a pure electric car, or if your circumstances dictate, a hybrid that spends most of its time in EV mode. Then have solar energy collection for your house in the form of solar electric panels, solar water heating panels, or whatever best suits you.

We are currently going through Brexit which is going to help us, as a country, stand on our own two feet again politically, but do you know how much energy we currently buy from other countries? No-one likes the idea of fracking but currently, we rely on other countries for our gas, and it is possible that they could turn off the taps. America has wanted to be self-sufficient with oil for decades, and I think they have achieved it. We, the UK could be self-sufficient with renewable energies if we, as a country, decided that it is what we want to achieve.

Let's all help our country along by doing the right thing.

Chapter 10 – 2019 mid-year update

I was fortunate again this year to go to Fully Charged Live at Silverstone. Some of the things that I saw there I have already covered in the book. The other things, plus anything else I have come across since my last re-write in October 2018 are in this chapter.

Peugeot e208 e3008
I was surprised and shocked to see a Peugeot at Fully Charged Live. So, being the inquisitive guy that I am, I went and asked some questions.

I like what they have done. Instead of redesigning the car from the ground up, or shoehorning in an EV kit, they have made whatever changes that were necessary to install the new power plant and batteries, but have retained the character of the car. The e208 and then a week later the e3008 were announced to the world in June 2019. The e208 will be available from as early as July or August 2019 with the e3008 being available early 2020. My local Peugeot dealer didn't know much more about it than that, but as I left my card, they will get back to me when more details are available, and the test cars arrive.

What I do know is that the same power unit is used in both cars. One thing that I was impressed with about the 3008, the petrol one, was that it only has a 1.2-litre engine. Like mine, it has a turbo, but that is just to bring up the power not particularly to super boost it. So, despite the large size of the car, it doesn't weigh a lot. Actually, they come out at 1.2 to 1.4 tonnes. Compare this to the 1.8 tonnes of my Nissan Murano, which is a similar size car. Obviously, this is what the petrol cars weights, the electric car will weigh more. According to google it come out at 1.4 tonnes, so a similar weight as I assume, the diesel engine car.

Stats wise, they are claiming 211 miles on a 54 kWh battery. Some people will look at this and still say it still does not have enough range. I'll tell you what three hours of driving is more than enough time for anyone. And these cars have a trick up their sleeves. They have the ability to charge at 100 kW. To put that into real-world stats, that is 0 to 80% in less than forty minutes. So drive for a couple or three hours. Grab a coffee and use the facilities, and you have enough juice to complete your journey. As long as you plug it in every time, you stop the car, will last longer than you can.

With 350 kW charges becoming more common, imagine what that will do to charging times when these cars can use them.

Renault Zoe test drive
This is a nice car. The only thing that I would say, having driven and been a passenger in a Nissan Leaf, is that there is not a lot of difference. While out on the test drive the rep from the car company told me that Renault owns something like 40% of Nissan. When I looked into this, apparently Nissan owns 15% of Renault. I have no idea how that works. He also mentioned that a lot of the Leaf parts are the same as the Zoe. That makes sense. To create a whole new system is expensive. If somewhere in your group, there is a company that has already created the parts, then it would be daft not to utilise them. It is like Peugeot using the same EV kit on both the e208 and the e3008.

So what is the Zoe like to drive? It is like a Leaf. Unfortunately, I am not a car journalist, so I didn't spot a lot of differences. If I was going to buy one of these cars, either a Leaf or a Zoe at a second-hand car place it would all come down to price and miles. I would maybe buy a Zoe so as to have a different car to my daughter, or I might get a Leaf because I know them. It really would be 50/50.

Don't be mistaken, it is a nice car. I don't know if it has the same acceleration as the Leaf but, although electric cars are known for their acceleration, just as in a petrol car, it uses up a lot of energy. I, unfortunately, have to say that I wasn't inspired by the acceleration on the Zoe. It felt like the Leaf in eco mode, which he said was turned off. Even in this mode, they are quite rapid. A lot better than my Dacia which turns into a snail in eco mode.

One of the reasons why my daughter got a Leaf instead of the Zoe a couple of years ago was because Renault was separating the price of the car from the battery. You leased the car for one price, then depending upon how many miles you thought that you were going to cover, determined the price of the lease of the battery, far too complicated. Nissan, two years ago said £137 a month all in.

I think that they have changed that system now. Car companies were not sure what to do as they didn't know how long the batteries were going to last. Now they realise that the batteries will in fact out last the cars.

So what else can I tell you about the Zoe? I think it is a sportier looking car than the Leaf. The early Leafs are great cars but do look a bit odd. I do like the newer ones. I'm sure that I saw something the other day about the Zoe getting a larger battery and bigger motor, I can't find anything about that today. This would make the sportier performance match the sporty looks. The current 40 kWh batteries are great, but a bit more range in a 50+ kWh battery is always good.

Tesla Model 3

A bit smaller, and from the looks of it a bit narrower than the Model S, the new Model 3 is a still a stunning car. I did get some pictures of it from the Live event, but they are not really book worthy. You can find far better ones on the internet. I'm not sure when I will get to drive one, but I will update this book as soon as I do. A friend at work has been asking about the Model 3. I said, unfortunately, there are no right-hand drive ones in the UK, and I don't know when they will be arriving. The following day there was a Fully Charged episode showing people in London picking up their right-hand drive model 3's.

The noticeable difference between the Model 3 and just about every other car on the road is that it has a seventeen-inch touch screen on the fascia between the two seats. Everything is controlled from there. I suppose one other feature that is worth mentioning is the rotary switches on the steering wheel. In my car, I have a different set of switches for temperature, volume and cruise speed. In the Model 3, it depends on what you have set as to what the rotary switch operates. It's a bit like the systems that now operate in aircraft. Instead of a switch having a single purpose, it just informs the computer of your decision.

Battery technology
Don't you hate it when the idea that is bouncing around your head is already being looked into? I watched a Fully Charged episode a few months, or a year ago when Robert went to see some Super Capacitors being used. The recharging speed of these is amazing. If memory serves me right, there are some busses that run on them. Unfortunately, Capacitors don't store electricity for very long. If they are topped up several times a day, they are happy. If you can ever call an electronic component happy...

Current car technology batteries charge great to 80% but then slow down and take literally hours to get to 100%.

So my thought was, why not combine the two technologies. To put it in simple terms, if you have 100 kWh of batteries, you need 20 kWh of capacitors. When you unplug your car when the main battery is showing 80%, the capacitors will be full and will trickle the power over to the main batteries. That way, charging times will be reduced even more.

Aircraft

If you were to take the powertrain and electrical gubbins out of a Leaf and put it into an ultra-light aircraft, you would have something that you could fly. People have added motors and batteries to primitive microlights, these are very successful. There are a couple of gliders/sailplanes that have the ability to self-launch on electricity. There is even the Pipistrel ALPHA Electro, which is an all-electric two-seat training aircraft. Unfortunately, at the moment, this aircraft doesn't have the flight endurance to complete all flight training. Some cross countries require a flight of a couple of hours plus reserves. I don't think it will be long before there are several aircraft of this type capable of the two to three hours that will be required.

I would personally love to learn to fly in one of these. In a conventional petrol engine Cessna, Piper etc. a lot of the complexities of the aircraft and the majority of the cost is for the engine. Electric ones are simpler to operate, cost a lot less to run and have little to no maintenance. Plus, using electricity, they are a lot better for the environment.

The cost of learning to fly today is say around £170 an hour, including an instructor at £25 an hour. Of the aircraft costs of £145, a third of that is fuel at around £48, another third will be for Engine fund, repairs and maintenance. The other third will be for the Airframe, i.e. the rest of the aircraft. Using those numbers, we will be looking at between £75 and £100 for a flight in electric aircraft. Even if we assume £100 an hour, that would save over £2,000 on flight training based on 45 hours. In addition, I have had flights cancelled due to engine problems or overrunning maintenance.

I was actually surprised to found out that there are a over a hundred electric powered aircraft projects out there.

Norway has declared that all short commercial flights in the 25 to 30 seat size whose frequent flights are 15 to 30 minutes are to be electric. They want the first aircraft to be operational by 2025. It is nice to see countries taking this issue, and technology, by the horns. The shortest commercial flight in the world is to one of the Scottish islands. It is part of a flight sequence, but I think the total flight time is still less than one hour for all three trips. They are looking to replace the current aircraft with an electric one in the very near future.

While it cannot currently be foreseen how long haul aircraft can be replaced with electric ones, there are certainly many ways that larger aircraft can be made more fuel and energy efficient. Firstly, a lot of commercial aircraft have two engines these days, one under each wing. They are required to have enough thrust that, should they have a bird strike that disables one engine on take-off, there is still sufficient thrust to be able to make a safe take-off, go around and landing. In the cruise, however, these engines are throttled back to around 60% to maintain height.

The problem is that, as far as I can remember from my training days, the slower a jet engine turns, the less efficient it is. Yes, it will use less fuel, but a smaller engine at higher RPMs will use even less fuel.

This could be overcome by using a smaller engine in the fin and motors on the wings. A second engine could be inside the fuselage like an oversized APU (Auxiliary Power Unit). The motors would be powered by the batteries, or if the fin engine had a problem, the APU could supply sufficient electrical power to enable to flight to proceed. This fin engine would basically provide 100% of the cruise power required.

Another area where fuel can be saved is in taxiing. Even on a typical flight as much as two tonnes of fuel can be used getting to the end of the runway. Sometimes aircraft have to wait behind others for their departure wasting more tonnes of fuel. This can result in engines being shut down, and at least one aircraft has crashed with a rushed startup. If the engines aren't started until just before entering the runway in everyday practice, then the checklists will be performed the same every time.

Motors in the main wheels and batteries stored in the fuselage would power the aircraft along the taxiway, they could also be used to help the aircraft accelerate from a standstill.

There are two other advantages to this system. A lot of tyre rubber is lost when the wheel touches the runway, and it spins up to speed. If the wheels could be spun up to the right speed just before touchdown them a lot of this tyre wear could be elevated. This could eventually reduce the weight of new tyres. Additionally, once on the ground, the wheel motors could be used to slow the aircraft down, this reduces the size of the brakes, and the regen recharges the batteries.

There is a motor glider in development that uses powered wheels for taxiing and take off assist and a jet engine for sustained thrust. The wheels can accelerate the aircraft to 50 mph on their own. At this speed, the jet engine has enough thrust to maintain the acceleration and climb the aircraft to soaring altitudes.

While the Jet engine would be presumed to run on Jet fuel, i.e. Kerosene, it can actually run on vegetable oil, therefore, making it totally suitable for this book.

ElectraFlyer-ULS

The ElectraFlyer is an ultralight sports aircraft that I think is awesome. As an ultralight, it has a two-hour duration, but as a light aircraft, it can stay aloft for up to four hours. Ultralights are an American standard where people can fly without medicals or licences, although training is advised. It's a bit like riding a motorbike on learner plates.

With single person operation and a 40 mph cruise speed, this could make for a delightful afternoon of gliding, soaring. Using thermals, ridge lift etc. could really extend the range of this aircraft if wanted.

Chapter 11, Finishing Words

Well, folks, that's about it. I just want to finish off by saying that I hope that you have enjoyed reading this small book. It is my intention in writing this to empower you with some more knowledge to help you see where the world is going, so that you are not surprised when things change. To give you some idea of where to start looking when making renewable energy and EV decisions in your life, and to hopefully get you excited as I am about this new technology and with it, doing our bit to help save our environment.

We won't be saving the planet. She can't really be hurt by humans and will fully recover after we have gone. But making the changes highlighted in this book we will at least lower our personal CO_2 and other emissions. Therefore doing our little bit to clean up the air that we breathe.

Don't forget that, I like you, are just a person learning about the next paradigm, and therefore all of my comments and advice are based on what I have learned.

Who knows, if I can help enough people to change their lives with my book, I might be able to afford to buy my dream electric car.

www.ingramcontent.com/pod-product-compliance
Lightning Source LLC
Chambersburg PA
CBHW020449220526
45464CB00002B/923